INDONESIAN HERITAGE

Plants

GROLIER INTERNATIONAL, INC.

EDITORIAL ADVISORY BOARD

Dr. Haryati Soebadio (Chair)
H.E. Minister Joop Avé
Dr. Edi Sedyawati
Soedarmadji Damais
Dr. Bambang Hidayat

Dr. S. Somadikarta
Dr. Aprilani Soegiarto
Dr. Taufik Abdullah
Paul Sochaczewski
Dr. John Miksic

SERIES EDITORIAL TEAM

Publisher
Didier Millet

Editorial Consultant
Peter Schoppert

Editorial Director
Timothy Auger

Senior Editor
Dr. Sian Jay

Editors
Dr. Julian Davison
Amita Sarwal

Assistant Editors
Goh Geok Yian
Tan Hwee Koon

Art Director
Tan Tat Ghee

Designers
Joseph G. Reganit
Norreha Bt. Sayuti
Nelani Jinadasa
Lee Woon Hong
Felicia Wong

Design Consultant
Michael A. C. Beech

Illustrators
Anuar Bin Abdul Rahim
Atang Fachruroji
Bruce Granquist
Duki Noermalla

Project Coordinator, Jakarta
Seti-Arti Kailola

Administrative Coordinators, Jakarta
Nunun Daradjatun
Tjetjep Suparman

Editorial Coordinator, Jakarta
Judi Achjadi

Editorial Assistant, Jakarta
Rafli Lindaryadi

Production Manager
Edmund Lam

Picture Researcher
Nicholas Tan

SPONSORS

This encyclopedia project was initiated and guided by the **Yayasan Dana Bakti**

with the support of the **Shangri-la Hotel**, *Jakarta*

It was also made possible thanks to the generous and enlightened support of the following companies

Sinar Mas Group
Bakrie Group
Bank Artha Graha
Satelindo
Telekomindo
Telekomunikasi Indonesia
Indobuildco
Indosat
Inti
Pasifik Satelit Nusantara
Plaza Indonesia Realty
Siemens Indonesia
WES Intratama Consortium
Wahana Tigamas Buana/AT&T
Konsorsium Promindo Ikat

Japanese Telecommunications Association
(NTT, Sumitomo, Fujitsu, Itochu, Kanematsu, KDD, Marubeni, Mitsui, Nisshoiwai, Tomen, Denki Kogyo, Fujikura, Furukawa, IKI, Infotama, JRC, Nesic, Bukaka, NEC, NNC, NTC, Sarana Ekacitra Indonesia, Tamura Denki, Voksel)

Nusa Telekomindo Internusa
Artha Telekomindo
Amalgam Indocorpora
Elektrindo Nusantara
Jatimas Fadjar Satryo Group
Komselindo
Citra Permata Sakti
Adcor Utama

© Editions Didier Millet, 1996

593, Havelock Road. # 02-01/02
Isetan Office Building
Singapore 169641
Ph. (65) 735-7990

Printed by Pt. Jayakarta Agung Offset, Jakarta

Published by **Buku Antar Bangsa** (Jakarta)
for **Grolier International, Inc**

Distributed exclusively by
PT. Widyadara
Wisma Anam
Jl. Teuku Cik Ditiro, No. 12
Jakarta, Indonesia 10350
Ph. (62-21) 310-6886, 310-1635

All Rights Reserved. No part of this book may be reproduced, stored in a retrieval system, or transmitted in any form or by any means without permission from the publisher.

INDONESIAN HERITAGE

Plants

VOLUME EDITORS

Dr. Tony Whitten *Ecologist, Asia Technical Division, The World Bank,*
and **Jane Whitten** *Ecologist*

VOLUME EDITORIAL TEAM

Tan Hwee Koon
Amita Sarwal
Editors

Joseph G. Reganit
Designer

David Stone
Dr. M. M. J. van Balgooy
Consultant Editor

AUTHORS

Elizabeth Bennett – *The Wildlife Conservation Society, Sarawak, Malaysia*
Jan Willem Taco Bottema – *ESCAP CGPRT Centre, Bogor, Indonesia*
Deni Bown – *Freelance writer, Milton Keynes*
Julian Caldecott – *Independent consultant, Wiltshire*
Josephine Camus – *The Natural History Museum, London.*
Martin Cheek – *Herbarium, Royal Botanic Gardens, Kew*
Janet Cochrane – *Freelance writer and eco-tourism consultant, UK*
Jill Cowley – *Herbarium, Royal Botanic Gardens Kew*
Phillip Cribb – *Orchid Herbarium, Royal Botanic Gardens, Kew*
John Dransfield – *Royal Botanic Gardens, Kew*
Soejatmi Dransfield – *Royal Botanic Gardens, Kew*
Alan Eddy – *The Natural History Museum, London*
Michiel Flach – *Tropical Crop Science, Wageningen Agricultural University, The Netherlands*
James J. Fox – *Research School of Pacific and Asian Studies, Australian National University*
David Galloway – *Consultant Lichenologist, New Zealand*
Wim Giesen – *Freelance Wetland Consultant, The Netherlands*
Ian Glover – *Reader, Southeast Asian Archaeology, University College, London*
Alison Hoare – *Department of Economic Botany, Royal Botanic Gardens, Kew*
Camilla Huxley – *Local wildlife conservationist, Oxford*
Matthew Jebb – *Trinity College, Dublin*
Don Kirkup – *Royal Botanic Gardens, Kew*
Anthony Lamb – *Kota Kinabalu, Sarawak, Malaysia*
David Mabberley – *Wadham College, University of Oxford*

L.J.G. van der Maesen – *Department of Plant Taxonomy, Wageningen Agricultural University, The Netherlands*
Genevieve Michon – *International Centre for Research in Agroforestry, (ICRAF), Bogor*
Kathryn Monk – *A consultant from the Zoological Society of London to the Leuser Development Programme, Medan, North Sumatara, Indonesia.*
Francis Ng – *Research Support, Centre for International Forest Research (CIFOR), Jakarta*
Roger Owen – *Freelance writer, London*
Caroline Pannell – *Freelanace biologist, rain forest ecologist and plant taxonomist, UK*
Oyvind Sandbukt – *Centre for Development and the Environment, University of Oslo*
Trond K. Schumacher – *Department of Biology, University of Oslo, Norway*
J. M. B. Smith – *Department of Geography and Planning, University of New England, Australia*
Mohamad Soerjani – *Centre for Research on Human Resources and the Environment, University of Indonesia*
Benjamin Stone – *Botanical Research Institute of Texas*
Tan Teck Koon – *Department of Botany, National University of Singapore*
Ian M. Turner – *Department of Botany, National University of Singapore*
Elizabeth Widjaja – *Herbarium Bogoriense, Indonesia*
Tim Whitmore – *Department of Geography, University of Cambridge*

GROLIER INTERNATIONAL, INC.

Contents

INTRODUCTION
Ian Turner

DIVERSITY OF INDONESIAN FLORA
Tony Whitten

Origins of Indonesia's Flora 10
Tim Whitmore & David Stone

Determinants of Vegetation 12
Tim Whitmore

Forest Dynamics 14
Tim Whitmore

Adaptations in Response to Diverse
Ecological Niches 16
Tim Whitmore

Fungi 18
Tan Teck Koon

Lichens 20
David Galloway

Bryophytes 22
Alan Eddy

Ferns 24
Josephine Camus

Plants Which Bear Seeds 26
David Mabberley

The World's Largest Flower 28
Deni Bown

VEGETATION TYPES AND BIOGEOGRAPHY
Tim Whitmore

Lowland Rain Forest Formations 32
Tim Whitmore

Freshwater Wetlands 34
Wim Giesen

Coastal Wetlands 36
Wim Giesen

Montane Forest 38
J M B Smith

Alpine Vegetation 40
J M B Smith

Monsoon Forest 42
Kathryn Monk

Savanna Forest and Grassland 44
Kathryn Monk

Secondary Forest 46
Tim Whitmore

PLANT–ANIMAL INTERACTIONS
Tony Whitten

Pollination by Animals 50
Ian Turner

Dispersal of Seeds and Fruits 52
Caroline Pannell

Mistletoes 54
Don Kirkup

Carnivorous Plants 56
Martin Cheek & Matthew Jebb

Plant Defences Against Herbivores 58
Elizabeth Bennett

Plants and Ants 60
Camilla Huxley

PEOPLE AND PLANTS: FOREST PRODUCE
Tony Whitten

Food from the Forest 64
Øyvind Sandbukt & David Stone

The Austronesian Botanical Idiom 66
James J. Fox

Plants Used in Medicine 68
Trond Schumacher

Resins 70
Tim Whitmore

Rattan 72
John Dransfield

Bamboo 74
Soejatmi Dransfield

Timber: Species and Kinds .... 76 *Tim Whitmore*	Sugarcane and Tobacco .... 112 *Tony Whitten*
Sustainable Use of Indonesian Timber .... 78 *Julian Caldecott*	Coffee, Tea and Cocoa .... 114 *Tony Whitten*
The Challenges of Indonesian Forestry .... 80 *Julian Caldecott*	Oil Palm and Rubber .... 116 *Tony Whitten*

PEOPLE AND PLANTS: CULTIVATION
Tony Whitten

Origins of Cultivation .... 84 *Ian Glover*	Timber and Softwood Plantations .... 118 *Tim Whitmore*
Rice and Rice Fields .... 86 *Roger Owen*	Orchids: An Introduction .... 120 *Phillip Cribb*
Indigenous Gardens: Re–inventing the Forest .... 88 *Geneviève Michon*	Cultivated Orchids .... 122 *Phillip Cribb*
Legumes: An Introduction .... 90 *L.J.G. van der Maesen*	Ornamental Plants .... 124 *Francis Ng*
Legumes: Food and Forage .... 92 *L.J.G. van der Maesen*	Making the Most of Weeds .... 126 *Mohamad Soerjani*
Root and Tuber Crops .... 94 *Jan Willem Taco Bottema*	Potential for development from Plant Genetic Variety .... 128 *Elizabeth A. Widjaja*

CONSERVATION
Tony Whitten

Fruits and Nuts of Indonesia .... 96 *Anthony Lamb*	National Parks and Other Reserves .... 132 *Janet Cochrane*
Palms: An Introduction .... 98 *John Dransfield*	The Role of Botanical Gardens .... 134 *Benjamin Stone*
Palms: Coconut and Areca Nut .... 100 *Alison Hoare*	Glossary 136
Palms: Sugar, Lontar and Nipa .... 102 *Alison Hoare*	Bibliography 139
Palms: Sago Cultivation .... 104 *Michiel Flach*	Index 140
Gingers .... 106 *Jill Cowley*	Photo Credits 144
Nutmegs and Cloves .... 108 *Tony Whitten*	
Pepper .... 110 *Tony Whitten*	

INDONESIA: A Botanical Bounty

Indonesia abounds in botanical riches. Nearly 10 per cent of the world's plant species are found here including some of the most beautiful and fascinating. The native species provide the framework of a series of ecosystems, ranging from sea to mountain top, that clothe the islands and shores of the Indonesian Archipelago in a vibrant and varied tapestry of forests and savannas.

The Important Role of Plants

Plants, through their ability to photosynthesize, provide all the energy available to the animals in an ecosystem and are thus irreplaceable. They are also important, in collaboration with microorganisms such as bacteria and fungi, in the perpetual cycling and recycling of important nutrients in the living community. Vegetation also protects the soil from erosion by rainfall and may help cool the local climate by evaporating large quantities of water into the air by the process of transpiration.

Human beings are no exception when it comes to animal dependence on plants. Plants have always been our major source of food, as well as an important providers of many other useful substances including medicines, fibres, building materials, dyes and cosmetics. The myriad cultures of Indonesia are united in their inclusion of plants in daily life as well as ritual and religion.

In this volume we explore the diversity of plants and vegetation to be found in Indonesia. From the snowy peaks and smouldering volcanic craters to the sandy shores and mangrove swamps, Indonesia's plants provide a vast spectrum of interest richly tinted by the beautiful and the bizarre, the perfumed and the putrid, the delicious and the deadly.

Indonesian Plants and Vegetation

The present volume presents an overview of the plants of Indonesia. The first half concentrates on the native plants and vegetation types of the Archipelago. Necessarily selective, the major plant groups and their habitats are featured.

Many different systems have been developed to classify the diversity of the botanical world. Supposedly 'undeveloped' human cultures generally divide plants into classes of edible and inedible. Plant taxonomists use a system that reflects the proposed evolutionary relationships of different plants groups.

Major branches of the evolutionary tree of plants are occupied by algae, bryophytes (liverworts, hornworts and mosses), pteridophytes (ferns and their allies), gymnosperms (the cycads, conifers, meninjaus and their kin) and the angiosperms (flowering plants). The fungi represent several evolutionary lines, not very close to the plants. The algae span a large area between the true plants, the animals and the microbes like fungi and bacteria. Lichens are living associations between fungi and algae. All these groups are well represented in the flora of Indonesia and are featured and illustrated in this volume. Some of the most famous botanical denizens are represented, including those floral giants: the titan arum *(Amorphophallus titanum)* and the rafflesias. The insectivorous pitcher plants had to be included too.

The different sorts of plants come together to form the community

A series of stamps were introduced in 1995 to create awareness for the flora and fauna of Indonesia. These include species such as sandalwood, eboni and sealing wax palm.

Painted motifs on natural textile used to make clothing for the people of Pulau Halmahera in the Maluku. The cloth is made from beaten bark.

«««Brightly coloured rhododendrons are a common sight in the mountains of Indonesia. The specimen illustrated, Rhododendron javanicum is obtained from a colour plate from the famous Flora of Java.

that makes up the particular vegetation found at any place. Tropical rain forest, covers much of Indonesia, and is the linking theme of much of the text. The rain forest is not a single entity, but varies considerably between sites. Indonesia is so large that the species composition of the forest at one end of the country is generally very different from that at the other, so while a patch of forest in Sumatra might look superficially very similar to another in Irian Jaya, the number of tree species in common would be very few indeed. The forest varies over smaller spatial scales as well. Factors such as soil type, altitude and rainfall all have some influence in determining which species thrive at a particular site.

The forest is not inhabited by plants alone, and one section of the volume is devoted to the variety of interactions that occur between plants and animals. These include pollination, seed dispersal, carnivory by plants and also herbivory by animals.

The fiery red ginger Tapeinocheilos pungens, *originates from the forest floor of Irian Jaya.*

Cili — doll with a typical silhouette of a girl with a slim hourglass body, with rounded breasts, long thin arms, great earplugs, and wearing an enormous headdress of flowers. These dolls which are used as offerings for agricultural ceremonies are made from the leaves of the sugar or coconut palm.

Oil palm, an important plantation crop in Indonesia covers about 1 million hectares in land area and originates from the rain forest of West Africa. Here workers tend the young oil palm in a nursery in Sumatra.

Plants and People in Indonesia

There are relatively few professional botanists in Indonesia, and much of the study of Indonesian plants and vegetation has been conducted by foreigners, mostly Europeans. Yet most Indonesians do not need scientists to tell them of the importance of plants to their daily lives. They know that their lives revolve around the gathering of plants from the forest, the planting and tending of crops in their gardens and fields, the transporting and selling of their fruit or rubber, or work in the plantations or logging camps. Indeed, it is difficult to think of a nation where plants and their products have played a more central role in the history, culture and economy than Indonesia.

The botanical riches of the Malay Archipelago have attracted many outsiders seeking trade, or empire building. Spices and precious woods have lured countless adventurers from near and far to the tropical islands that today are Indonesia. The quirk of nature that endowed the Archipelago with nutmeg, cloves and pepper; sandalwood, eagle-wood and camphor; meranti, keruing and rattan, has had a profound influence on the historic, cultural and economic ties of the Indonesian people with places as far away as China, India, Arabia and Europe.

The exploitation of wild plants is the first focus of the second half of the volume, including collection of food, medicines, rattan, bamboo and timber. This leads on to a consideration of cultivated plants, including such important crops as rice, legumes, palms, gingers, spices and ornamental plants such as orchids.

Plants in the Future

We document the many uses people in Indonesia make of the plants they collect from the forests or grow in their gardens or fields. Market stalls, temple altars and dining tables bear witness to the diversity of plants to be found in Indonesia. Yet probably only a minority of species are used in any way, and the secrets of many species are known to but a few.

Plant and animal species have been likened to 'uncut diamonds'. When dug out of the ground a diamond looks, to the untrained eye, just like any other stone. Yet after skilful treatment by the gem cutter the diamond realises its potential as a priceless jewel. Many wild plants in Indonesia may have valuable properties we know not of. The enormous diversity of plants could easily include those with the power to counter diseases today considered incurable, or with genetic traits such as pest or pathogen resistance that could be transferred, at a great advantage, to our crop plants. We need to gain the skills of the diamond cutter to gather the resources on offer. The danger we face at present is that by destroying the natural vegetation of the world, including the forests, swamps and grasslands of Indonesia, we will discard some of the uncut diamonds.

Top (from left): The diverse features of the rain forest: Bright red fruit of a climbing vine. Fruits produced from the main trunk of a tree. Lichens in montane forest. Natural patterns on a palm leaf.

Bottom: A collection of Indonesia's orchids
- **A** Dendrobium taurinum
- **B** Phalaenopsis amabilis
- **C** Paphiopedilum praestans
- **D** Vanda sp.
- **E** Coelogyne pandurata
- **F** Paphiopedilum urbanianum

DIVERSITY OF INDONESIAN FLORA

Indonesia is botanically one of the richest countries in the world. This is partly due to its great size, but it also reflects the legacy of a complex geological history, the existence of many islands on which endemic species have developed, and a near perfect climate for plant growth in many parts of the Archipelago. Indonesia stretches across 5,100 kilometres and encompasses a very wide range of habitats, including permanent swamps, wet and dry forests, grasslands, semi-deserts and even a glacier. With this wealth of habitats comes an equally rich diversity of plant species.

All the major families of plants are well represented in Indonesia, but our knowledge of them varies enormously. Flowering plants, being popular subjects of study, are quite well documented, with few new species still being found. At the other end of the scale, however, there are the lichens — composites of algae and fungi — which are not at all well known in Indonesia, and there are doubtless many new species awaiting discovery.

The best known fungi are the conspicuous varieties, with estimates of the total number of species found in Indonesia ranging from 4,250 to 12,000, out of an estimated world total of 47,000 species. There are also around 3,000 species of mosses and liverworts in Indonesia, the world total being some 15,000 species. And there are probably some 4,000 species of fern, which represents perhaps a quarter of the entire global complement. The biggest group, however, comprises seeding plants, with 20,000 species representing eight per cent of the world total. They range in size from the minute floating *Wolffia*, just 0.5 millimetres across, with even smaller flowers, to the enormous rafflesia flower, which grows to nearly one metre across. The largest family of all is the orchid family, with some 30,000 species; some other families are represented by just a single species.

The distribution of plant species in Indonesia is far from homogenous. The richest areas are the primary lowland rain forests of Kalimantan, where 34 per cent of the 10,000 seeding species are endemic, that is to say they are found nowhere else in the world. This includes 158 out of 267 tree species of the commercially-important dipterocarp timber family. Sumatra and Irian Jaya are also very rich in species, but the rain forests of Java, Sulawesi, Maluku and the Lesser Sundas are relatively poor in this respect, though by no means impoverished when compared with many other parts of the world.

Origins of Indonesia's Flora

Indonesia's plant life, the species present as well as their distribution, is strongly influenced by the physical and topological characteristics of the Archipelago. These features have their origins in the ancient geological history of the region. Natural selection has also influenced the selection and survival of plant species that have arrived from various parts of the world, to form the unique assemblages of organisms which now characterise the floral communities of Indonesia.

GEOLOGICAL HISTORY OF THE REGION OF WALLACE'S LINE
While important land movements took place from 200 to about 50 million years ago, the two halves of the Malay Archipelago remained apart. From 53 million years ago, Australia & New Guinea moved north into the Tethys Ocean, converging on Southeast Asia. As late as 10 million years ago, extensive dry land was exposed between Australia and Celebes, providing a possible route for animal and plant migration. From left: Details of the continental movements accounting for the Gondwanic and Laurasian elements of the Malay Archipelago. **120 Million years**

The last phase of the glancing collision between Gondwana and Laurasia (early Miocene epoch) resulted in the intussusception of the Banda arc at Sulawesi by New Guinea.

Clockwise: The crater of Anak Krakatau, a site of constant activity where recolonisation studies are conducted. Germinating Barringtonia *seedlings washed ashore.* Saccharum spontaneum, *a pioneer species establishing itself on the volcanic debris of Anak Krakatau.*

Geological Changes

The land masses which make up the Indonesian Archipelago have been moving for millions of years, although at an imperceptible rate. The Archipelago, as we see it today, came into existence some 15 million years ago during the mid-Miocene epoch of the Tertiary era. It is the result of a collision between the two ancient continents of Laurasia and Gondwana, whose submerged forms can still be discerned beneath the shallow waters of the Sunda and Sahul continental shelves.

Laurasia and Gondwana were, themselves, once part of a single supercontinent, Pangaea, which split in two some 180 million years ago. Subsequently, the northern edge of Gondwana began to break up and drift northwards towards Laurasia which, at that time, extended southeastwards to as far as present-day Sulawesi. These Gondwanan fragments eventually collided with the Laurasian landmass, resulting in the present arrangement of islands in the Archipelago.

Continental drift has also been responsible for the formation of many of Indonesia's mountain ranges, as collisions between land masses created major disturbances of the Earth's crust. The great arc of volcanoes which sweeps southwards and eastwards through Sumatra, Java and the Lesser Sunda Islands is located along a plate margin where part of the Earth's surface lying beneath the Indian Ocean is being pushed under the Sunda shelf.

Different Continents: Separate Lives

Today's patterns of plant and animal distribution in Indonesia are a reflection of the region's complex geological history. Quite different animals evolved in Laurasia and Gondwana following the break up of Pangaea. Thus, the wildlife of western Indonesia differs remarkably from those islands in the east. A boundary between these faunas was first identified by the famous Victorian naturalist Sir Alfred Russel Wallace, who proposed that a natural barrier separated the islands of Bali and Lombok, and that this boundary extended northwards to separate Borneo from Sulawesi. The demarcation for plants by Wallace's Line is not so clear, but many groups do show a western (Laurasian) or an eastern (Gondwanan) concentration. The timber tree family Dipterocarpaceae, for example, which is of such importance in western Indonesia, is scarcely in evidence east of Wallace's Line: Borneo has 267 dipterocarp species, but Sulawesi only seven.

The division of plant and animal ranges into distinct geographical units is not, however, entirely clear cut. Long before the actual collision between Laurasia and Gondwana, fragments of the latter had been breaking off what is today northern Australia and drifting northwards across the Tethys Ocean. Geologists are still discovering these Gondwanan fragments — so-called terranes — embedded in the lands which today form the western part of the Archipelago, but which were once believed to be entirely Laurasian. The northwest corner of Borneo is one such example, and it seems likely that these fragments may have carried Gondwanan species to Laurasia prior to the collision of these ancient land masses 15 million years ago, though concrete evidence for this is hard to find.

Palaeoclimate and Changing Sea Levels

We live today in a relatively warm part of the Earth's history, known as an interglacial period. This is characterised by the polar ice caps being quite high. During past glacial periods, however, the lowland climate of the humid tropics was somewhat cooler, slightly drier and more seasonal. With much more water locked up in the polar ice caps, sea levels were up to 180 metres lower than today, while sea surface temperatures in the tropics were some two degrees Celsius cooler. Lowered sea levels meant that both the Sunda and Sahul continental shelves were exposed, so that Sumatra, Java, the Malay Peninsula and Borneo were all joined together by dry land, while New Guinea, Australia and their neighbouring islands were also united as a single landmass.

The present snowline on Indonesia's mountains

DIVERSITY OF INDONESIAN FLORA

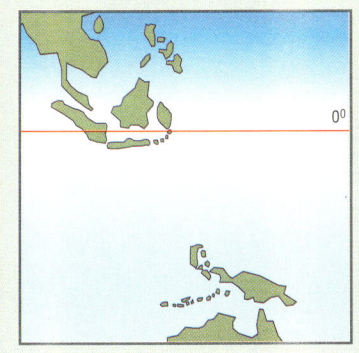
80 Million years

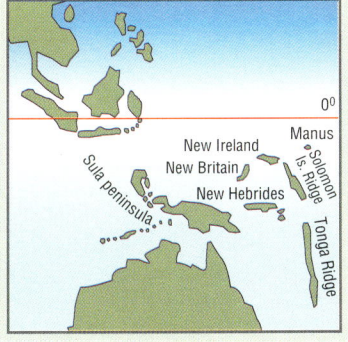
40 Million years

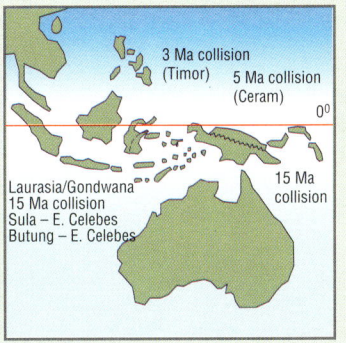
20 Million years

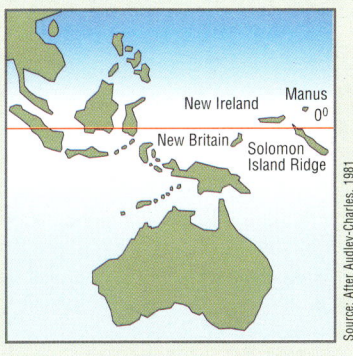
10 Million years

occurs at very high altitudes, while the vegetation zones below it are all broad. In the past, however, during glacial periods, the vegetation zones on tropical mountains were both lower and more compressed, with the forest limit on the mountains of New Guinea occurring at about 2,500 metres — some 1,000 metres below the present tree line. At the same time, lowland rain forests were far less extensive than they are today, while seasonal forests were much more widespread.

The height of the last glacial period was about 18,000 years ago after which the climate rapidly warmed and sea levels rose, until the beginning of the Holocene, some 10,000 years ago, when a climate rather similar to present conditions began. There have been many glacial periods over the past two million years, lasting far longer than today's inter–glacial period. In fact, the present climate and, hence, the wide extent of tropical rain forests is rather atypical.

Volcanic Upheavals

Natural forces at work within Indonesia have also had a noticeable impact on the landscape and resulting vegetation patterns. Massive volcanic eruptions have destroyed whole communities of plants and animals, creating huge areas of barren, volcanic rock which remained for centuries as natural barriers to the re–establishment of life. The vast calderas of Lake Toba in North Sumatra, Tengger in East Java and Batur in Bali, give some indication of the devastation and climatic changes that must have accompanied their catastrophic formation. But given enough time, plants and animals can recolonise most of these zones, as witnessed by the revegetation on islands such as Anak Krakatau which arose from the sea in 1930, presenting scientists with a rare opportunity to study the development of plant and animal communities on a barren landscape. Today, some 60 years after this island first appeared, thickets of casuarina and wild sugar cane are just some of the plants which have successfully become established.

Plant Ranges and Species Richness

The distribution of Indonesia's flora reflects the region's complex geological and climatic history. Some Indonesian palm species, and a few members of the Dipterocarpaceae family, are also found in the Seychelles, on the far side of the Indian Ocean, which betrays their common and ancient Gondwanan ancestry. Other plant species, such as the betel nut palm (*Areca catechu*), have concentrations at both the western and eastern ends of the Archipelago. These bicentric groups are believed to be plants of Gondwanan origin which arrived in Indonesia by the south Asian route described above, and also via Australia. Another factor which explains the presence of some species has been the deliberate, or accidental, introduction of plants by humans.

The rain forest flora of Indonesia is exceedingly rich in species. A survey of a 1.6 hectare plot at Wanariset in southeast Kalimantan revealed 240 species of trees with a diameter greater than 10 centimetres. The reason for such high levels of diversity are a continuing puzzle for botanists. One contributory factor is the high incidence of local endemic species, which may include species whose total world range is limited to just a few square kilometres. This local endemism, together with the co–occurrence of huge numbers of widely ranging species, has contributed to high levels of species richness in small plots: recent research has shown that areas a few kilometres across may contain from one–third to half of the known tree flora of individual islands. In such circumstances most species are inevitably found at low density.

From top left, anti–clockwise: Ipomoea pes–caprae, *a beach species widely distributed from East Africa to Polynesia.* Lobelia sumatrana, *a common herb on Indonesian mountains belonging to a genus familiar to the northern hemisphere.* Potentilla borneensis, *belongs to a genus with a northern origin.* She–oak *(*Casuarina junghuhniana*), a Gondwanic plant now widespread in the Archipelago.*

*The conifer (*Podocarpus neriifolius*) is Gondwanan in origin but now has a wide distribution.*

Determinants of Vegetation

The Indonesian Archipelago, together with the Malay Peninsula, New Guinea and the islands of the Bismark Archipelago constitutes the botanical region of Malesia. Such a vast area, inevitably contains a wide range of environments and different types of vegetation. These are classified forest formations, each one occupying a distinctive habitat.

»»*Nypa fruticans dominates the landscape in the brackish waters within the boundaries of Way Kambas in Sumatra, one of the oldest nature reserve in Indonesia.*

Forest Formations

No two forests are the same, and even within a single forest there will be considerable variation from place to place in forest stature and structure, as well as the mixture of species represented. Forest formations are often defined by their structural composition and their physiognomy — a term that refers to specific floral characteristics such as crown formation, the size, shape and texture of leaves, the occurrence of cauliflory and ramiflory (flowers which grow on the trunks and on the main boughs of trees), and whether or not the main trunk of the tree is supported by buttress roots. Structural features of the trees incorporate canopy height, the occurrence of 'emergents' (trees that rise above the main canopy), and the abundance and nature of climbers and epiphytes.

Within a given area, the main factors determining the type of forest are regional climate, elevation, whether the site is swampy or not — whether the water is derived from rainfall or underground sources — the structure and composition of the underlying rock and the type of soil developed from this. There are usually very distinct and clearly defined boundaries between major soil types and, therefore, also between forest formations. Within each formation, however, there are usually smaller, more gradual changes in soil structure and quality from one location to another. Such changes will also be reflected by the forest composition. Such variations provide a constant challenge for forest ecologists who seek to detect and understand such gradients.

CHANGES WITH ALTITUDE

The composition and physiognomy of forests also changes with increasing elevation. The most obvious physical difference is that the higher one climbs, the colder it becomes; the rate of temperature decrease being about six degrees per 1,000 metres. The tops of the highest mountains in Irian Jaya, which reach nearly 5,000 metres, have a permanent snow cap, while those in Sumatra, Sulawesi and Java, which reach 3,450 to 3,800 metres, experience occasional frosts. Rainfall at first increases with elevation but then declines, so that the uppermost parts of very high mountains are drier than the middle slopes, as recent studies of Gunung Kerinci in Sumatra have shown.

As one ascends a mountain, anywhere in the wet tropics, there is a sequence of forest formations: lowland, lower montane, upper montane and sub-alpine rain forest. In most instances the boundaries between different forest formations are not sharply defined, reflecting gradual changes in the environment. Sometimes, however, the change from lower to upper montane rain forest is rather abrupt because it lies at an elevation where rising, cooling air becomes supersaturated with moisture and condenses to form clouds.

Effects of Climate

Seasonal rainfall patterns play an important role in determining forest formations. Rain forests occur where there is either no dry season, or a short and variable one of about two months. Scientists define a dry month as one with 60 millimetres rainfall or less. Stronger seasonal climates support a group of forest formations known collectively as monsoon or seasonal forests. In Indonesia, as throughout the tropics, the latter have been extensively degraded by humans, and today, have been replaced by open woodlands or treeless grasslands: fire and grazing have been important factors in the development and maintenance of these second climate ecosystems.

Moisture

Forest formations are also determined by the moisture content of the soil and can be classified according to whether the ground is generally dry or whether it is permanently or periodically wet.

Dryland Formations

The most extensive dry land forest, called lowland evergreen rain forest, occur on what are known as 'zonal' soils, which are red to yellow, nutrient-poor, clay-rich soils. They are species-rich and once covered great areas of the Archipelago. Agriculture has reduced the lowland evergreen rain forest considerably, with major blocks now only remaining in Irian Jaya and Kalimantan and nearly all have now been logged, thus altering their structure.

These rain forests are not the only forest formation represented in Indonesia. The mineral conditions of limestone and ultra-basic (igneous) rocks support distinctive forest formations. Limestone is extremely porous and thus unable to retain much water. As a result, large trees are unable to survive and the resulting flora is characterised by small trees and often a rich herb flora.

The most frequent vegetation type of ultrabasic rocks, rich in aluminium, magnesium, iron and heavy metals, is also dominated by small trees, many of which belong to the myrtle family (Myrtaceae). Unlike these habitats, white podzolised sandy soils in many parts of Kalimantan support a highly distinctive formation known as heath forest. Locally known as *kerangas*, which means 'land which will not grow rice', the soil base is easily degraded once the forest cover has been removed by loss of humus and clay.

Wetland Formations

Wetlands also have distinctive soils and corresponding forest formations of which three main types occur in Indonesia: freshwater swamp forest, peat forest and mangrove. The freshwater swamp formations occurs where the land is permanently or periodically flooded. These are still very extensive in central Irian Jaya but have mostly disappeared from western Indonesia where this type of land has been drained for rice cultivation and other forms of agriculture.

Peat swamp forest, on the other hand, occurs extensively in eastern Sumatra and Kalimantan, as well as Irian Jaya. Peat is formed by an accumulation of waterlogged leaf litter and fallen tree trunks which fail to decompose in anoxic waters that exclude bacteria and aerobic organisms of decay. Indonesia's peat swamps were laid down after the last Ice Age, when the sea level rose and flooded coastal areas. In some cases they may extend over tens of kilometres in diameter and reach up to 20 metres in depth. Peat soils are extremely infertile in that, unlike freshwater swamps, plant mineral nutrients come almost entirely from the rain. Shallow peat soils less than one metre deep are suitable for agriculture, but deep peat is best left as forest.

In contrast to swamp forests, mangroves grow along coastlines and river esturies where salt water is present in high concentrations. These specialised trees have not only succeeded in overcoming the problems of living on an inhospitable substrate by developing aerial roots with pores that allow gaseous exchange, and have also developed adaptations to overcome saline soil. Mangroves play an important ecological role in stabilising coastlines and providing nutrient-rich, calm waters that act as important nursery zones for many fish and crustaceans.

Clouds gathering at an altitude of 1,200 metres on Gunung Baluran near the northeast tip of Java. Small mountains such as this commonly have a low altitude cap of motane rain forest.

Both the structure and composition of the vegetation changes as one ascends the mountain:
A *Mangroves, found on the clayey sheltered coasts.*
B *Lowland rain forest and hill forest — a species-rich formation with characteristically tall canopy and emergents up to 40 to 60 metres, a tightly closed lower canopy and sparse undergrowth.*
C *Lower montane forest dominated by forest with shorter trees.*
D *Upper montane forest is characterised with a single layer of trees with smaller stature and gnarled limbs.*
E *Sub-alpine and alpine.*
F *Alpine climax vegetation.*

Forest Dynamics

All forests are dynamic ecosystems which are, in a continual state of flux. Patches of mature forest are punctuated by gaps in the canopy where seedlings and young trees grow. Together they form a mosaic, the gaps in the canopy constituting areas of so-called 'building forest' which will, in time, reach maturity as the forest regenerates. Understanding forest dynamics is imperative for the sustainable harvest of forest products.

Calophyllum sp. *seedling in the sparse undergrowth of the undisturbed lowland rainforest of Gunung Palung National Park in West Kalimantan.*

CYCLE OF INORGANIC NUTRIENT IN A RAIN FOREST

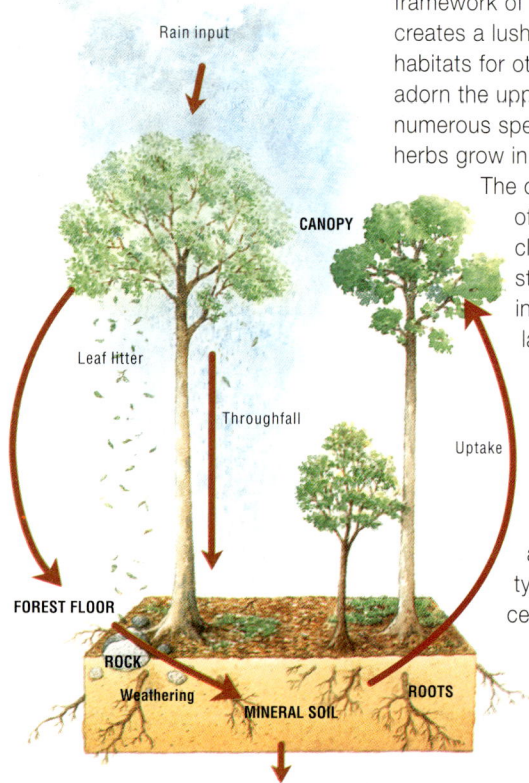

A Mosaic of Plants

In western Indonesia most gaps in the canopy are small, being formed when one or a few adjacent trees fall over or die. In this respect, the consequences of the disruption are relatively small and localised. In eastern Indonesia, however, where cataclysmic events such as earthquakes and landslides on the steep and young mountains are more common, the forest is periodically destroyed over areas reaching several square kilometres. Here, species whose regeneration is favoured by the harsh conditions of large openings are abundant.

In addition to these structural differences, the floristic composition of the forest may vary from one place to another because different tree species have different habitat preferences. Some species, for example, are especially common along rivers or on valley alluvium, while others predominate on ridge crests. Scientists are still trying to decipher and understand these preferences which probably emerge at the stage of seedling establishment. Changes in floristic composition tend to manifest themselves subtly: there are few sharp boundaries within the various forest formations.

A Myriad of Niches

Rain forests are complex ecosystems. The framework of a great number and variety of trees creates a lush canopy and provides a variety of habitats for other plants. Climbers and epiphytes adorn the upper levels, while on the forest floor numerous species of palm, ferns, wild bananas and herbs grow in the dappled shade of the larger trees.

The canopy is also home to a huge number of animals and, as with epiphytes and climbers, there is a high degree of stratification, some species preferring to inhabit the dark and relatively cool lower layers, and others the hotter, dryer and brighter upper levels.

There are close links between the plants and animals that comprise rain forest ecosystems. Nearly all plants are pollinated by animals and many species have evolved whose flowers are specially adapted to a particular type of animal as pollinator, for example certain species of bats and moths that have long tongues to reach the nectar supplies of long-flowered plants, long-beaked birds that may fill a similar diurnal role as well as many insects such as butterflies or bees that may be attracted to a few plant species. Other flowers, however, are relatively unspecialised and may be pollinated by any one of a number of animal species.

Animals play an instrumental role in the dispersal of fruit, and again, various strategies have evolved. Strong smelling fruits are especially attractive to mammals: tigers, elephants and orang utan, for example, have a notorious passion for durians. Birds, on the other hand, are attracted by brightly coloured fruits or seeds and these constitute the primary food source for frugivorous species such as pigeons and hornbills.

Effects of Logging

Logging forests for commercial purposes has developed into a major business. Felled trees create gaps in the forest canopy similar to those created naturally as old trees die and fall, often bringing down several others with them and damaging their neighbours. In both instances, the gap in the canopy is refilled by new growth from the forest floor. There are, however, some major differences between the two events. In the case of timber extraction, the removal of logs necessitates the building of tracks and access

DIVERSITY OF INDONESIAN FLORA

FOREST DYNAMICS IN A SECTION OF A PRIMARY RAIN FOREST

The forest growth cycle of the canopy is divided into three distinct phases: the gap phase — created by the death of a mature individual or a group of trees, the building phase — space created is rapidly filled up by secondary forest species and growing seedlings of the primary species, and the mature phase — when canopy is closed once again by the mature climax species. The forest is in a constant state of dynamic equilibrium, switching between phases.

FOREST STRATIFICATION:

The classical view of layering in a lowland rain forest is as follows:

A The highest layer consisting of isolated or grouped emergents
B The main canopy — a continuous layer
C A lower storey of trees directly below the canopy layer
D A layer of woody treelets
E Forest herbs and small seedlings

A logging road cutting across the virgin rain forest, opening the tightly closed forest structure. This facilitates the intrusion of urban growth by paving the way for the building of new settlements.

Forest Conservation

It is completely untrue, despite the rhetoric of environmental activists that logging destroys a forest. Certainly the structure is altered with the creation of more gap and building phase forest, but all the plant species are still present and the natural forest structure will eventually recover. Furthermore, research has shown that even the large arboreal animals continue to inhabit logged rain forest, exploiting the remaining canopy for habitat and food.

Nevertheless, it is important to minimise the amount of the forest floor destroyed by extraction tracks and to take steps to reduce erosion on them. The main logging roads, which will be needed for future harvesting operations, must also be protected from erosion while the encroachment of farmers, whose small holdings constitute perhaps the most serious threat to the future of the forest.

roads. Here the forest floor is scraped away and the use of heavy machinery compacts the subsoil; erosion causes gullies to form and water runs away more swiftly than would otherwise be the case. Even so, the forest will eventually reclaim these sites, but the process is much slower than when the forest floor is left more-or-less intact. Careful control over the direction of felling of trees and the planning of roads, can reduce the damage considerably.

Complete elimination of the forest by shifting cultivation, on the other hand, totally destroys the ecosystem. Thus it is essential to stop farmers from moving in along logging roads which have made previously remote areas readily accessible.

««Looking up at the canopy layer of the primary forest from below. As observed, most of the light rays are cut off from the bottom layers as light filters down the various strata of the three-layered structure. This accounts for the various strategies adopted by diverse plant forms in the intense competition to get to the top.

Adaptations in Response to Diverse Ecological Niches

Tropical rain forests present an almost overwhelming sense of abundance and luxuriance. This impression is attributable not only to the species richness of these ecosystems, but also to the remarkable diversity of plant forms in terms of their size, shape, growth pattern and habit.

The acorn-shaped red and yellow Balanophora elongata, *a root parasite found on Gunung Gede in Cibodas, West Java.*

*Bright red petals of the ginger (*Nicolaia solaris*) seen on the forest floor of Gunung Gede in West Java.*

Cauliflory, a common sight in the tropical rain forest, as exhibited by a member of the family Moraceae in Seram, Maluku.

The Fabric of a Forest

Trees are by far the most common life form in tropical rain forests. They come in many shapes and sizes, from minuscule forest floor treelets which may reach a height of just one to two metres, to enormous 80 metre giants, known as 'emergents', that rise above the surrounding forest canopy.

Most seedlings and young trees have a single stem with lateral branches, and grow upwards from an apical bud. This so-called monopodial construction subsequently changes to a sympodial construction as the tree matures. At this point the single apical bud is replaced by several growth points, and a crown of large limbs and numerous leading shoots develops. Emergents and those trees which form the top of the continuous forest canopy are mostly sympodial in construction, with their widespread crowns that are broader than they are deep. Lower down, within the canopy, the monopodial form is more commonly seen, and some species never lose it. Some examples are the nutmeg family (Myristicaceae), ebonies (*Diospyros*) and mangosteen (*Garcinia*).

Climbing Plants

Another conspicuous feature of most lowland rain forests is the presence of big woody climbers, or lianas. These are plants that are rooted in the ground, but have a long, fairly slender, often sinuous, vine-like stem, and a crown that is sometimes as large as that of a big tree, high in the forest canopy. The stems of lianas are often twisted, convoluted or buckled into strange and remarkable configurations.

The structure of liana stems is very different to that of trees, in that the water-conducting vessels in the wood are usually very large in diameter and are clearly visible to the naked eye. The stem of some species, when cut, provides drinking water for the thirsty traveller, others however, have poisonous sap.

The microclimate of the forest changes from the dark humid ground level, upwards towards the bright drier canopy top and different bole climber species grow along this gradient.

A second group called the bole climbers, attach themselves to neighbouring tree trunks. In some forests one finds individual trees which have been completely enveloped by the leaves and stem of these climbers. Important members of this group include the aroids, which belong to the Araceae family, and climbing pandans (*Freycinetia* spp.).

Climbing palms, or rattans, constitute another major group of climbers, found abundantly throughout the Malesia floral region. Slender, whip-like growths which are either modified infloresences or elongated leaftips and are equipped with hundreds of reflexed or recurved thorns attach themselves to any suitable means of support, while the lower stems of these plants are also covered by vicious, spiny leaf sheaths, which can deal a serious wound to the unwary passer-by.

Herbs Aplenty

A characteristic feature of tropical rain forests is the presence of giant herbs, which are quite different to anything found outside the tropics. Many belong to a group of related families of monocotyledons — the gingers (Zingiberaceae), bananas (Musaceae) and Marantaceae. These are typically found in disturbed, moist places, occurring naturally along valleys and nowadays along damp logging tracks. The forest floor is also home to a wealth of herb species. These are often patchy in occurrence and are particularly common on earth banks and rocky outcrops where leaf litter does not accumulate. Herbs are more frequently encountered in the mountains than the lowlands. They are particularly well-developed in forest growing over limestone where many spectacular species of Gesneriaceae are found.

Decorative Epiphytes

Epiphytes are another life form which occur in great abundance in tropical rain forests. They are not rooted in the ground but are attached to other plants, typically trees, which provide them with support. Epiphytes are not, however, parasitic and like other green plants must satisfy their energy requirements through photosynthesis.

In the dark, damp, lower part of the forest canopy, tree boles may harbour many species of epiphyte mixed with bole climbers. These may include a group known as hemi-epiphytes which start life rooted in the ground, grow up a trunk, and then die off behind. Epiphytes of the lower canopy are able to thrive in shaded places where there is little light. They have delicate thin leaves and need moist conditions. Another group of epiphytes, the sun-loving species which have thicker, fleshy leaves, are found higher up in the canopy and among the tree crowns. These species are well adapted to resist desiccation in the hot, bright, dry conditions found higher up, at the top of the forest. Most sun epiphytes have the same kind of specialised photosynthesis found in plants of semi-arid places and deserts. A number of orchids, for example, are epiphytic as are numerous ferns. Both these groups have species that are both shade and sun inhabiting.

Plants Without Chlorophyll

Saprophytes are plants which have no chlorophyll and are dependent on decaying organic matter for their nutrition. They are often whitish in colour; most are either tiny forest floor herbs or fungi.

Parasitic Plants

Parasites are plants which depend on others for their nutrition. The most spectacular parasitic plants in the world occur in the Malay–Indonesian Archipelago. They belong to the genus *Rafflesia*, named after their discoverer, the Englishman Sir Stamford Raffles, founder of Singapore and an early 19th century governor of Java. *Rafflesia* flowers grow on the surface of lianas of the vine genus *Tetrastigma*. The *Rafflesia* plant has no leaves, only flowers, plus hyphae which permeate the host tissues, from which they draw water and nutrients. *Rafflesia* flowers vary from 40 centimetres across to the gargantuan *Rafflesia arnoldii* of Sumatra which is over one metre across. The flower has a strong offensive smell and is pollinated by flies. Like other fly–pollinated flowers it is blotched purplish, red and white.

Besides the truly parasitic plants there is a large and important group called hemi–parasites, which is mainly made up of mistletoes (Loranthaceae). Although these tap a host tree for water and mineral nutrients, they do process chlorophyll and so are able to photosynthesize and make their own sugars. Hemi–parasites commonly develop as dense bushes, one to two metres across, high in the crowns of big forest trees. Mistletoes typically have long, tubular, brightly coloured flowers which produce watery nectar. This accounts why they are usually pollinated by birds.

The massive rafflesia: *its size in relation to a human being.*

«*Strangler fig in the lowland rain forest of the Cyclops Reserve near Jayapura in Irian Jaya.*

«*A magnificent display of the bird's-nest fern* (Asplenium nidus), *commonly found 'perched' on the trunk of a mature tree.*

Supported on a tree of the primary forest, the twisting stem of a liana demonstrates its ability to climb up the closed canopy. This evolves in response to an intense competition for light. This example is in West Kalimantan.

Fungi

Fungi are a diverse and common group in Indonesia, although only certain species are conspicuous. For taxonomic purposes fungi are generally placed in their own kingdom, distinct from green plants. They differ because they do not contain chlorophyll, and manufacture their food mainly by chemically digesting plant and animal matter.

Puffball fungi (Lycoperdon perlatum) *discharging spores from their fruiting bodies.*

Where Fungi Occur

Fungi are found in soil, water, vegetation, animal bodies, food and even on buildings. Their reproductive units, known as spores, are even present in the air we breathe. Despite being widespread most are very small and inconspicuous. Fungi grow and proliferate best in warm and humid conditions. It is thus not surprising that in Indonesia, with prevailing high humidity and temperature, there are thousands of species. To date, more than 70,000 species of fungi have been described worldwide, although it is estimated that there could actually be between 100,000 and 250,000. A significant proportion of these could be expected to be found in the little-explored forests and other ecosystems of Indonesia.

Basidiomycetes

To most people, fungi are synonymous with the large reproductive and spore-bearing fruit bodies, although these represent just one class of fungi, the Basidiomycetes. Fruit bodies exist in a wide variety of shapes, colours and sizes. Most people have noticed Basidiomycetes in the form of mushrooms appearing by paths and gardens.

The more common mushrooms include species of *Lepiota* and *Macrolepiota*. These are gregarious white mushrooms with brown scales on the caps. Another frequently found, but less conspicuous, species is *Coprinus plicatilis*. *Coprinus* are small, delicate mushrooms which decay rapidly into an inky mess, thus earning the common name of 'ink caps'.

A rather unpleasant name is given to an exquisite wayside fungus with a white, yellow or orange veil — the stinkhorn fungus (*Dictyophora indusiata*) which attracts flies to its spore-bearing tip. A similar, but less common species that often forms extensive

mats on litter of *Acacia auriculiformis*, is *Scytinopogon angulisporus*. Unusual names are also given to other fungi including the jelly or ear fungi (*Auricularia* spp.) which occur on rotting logs and branches, and the coral fungi (*Ramaria* spp.).

Since Basidiomycetes are decomposers of wood, a greater abundance and diversity of these fungi is found in forests where there are quantities of fallen trunks, branches and leaves. These fungi play an important role in the decomposition of dead plant matter and the recycling of carbon. Such environments support many common gilled fungi such as *Marasmius*, *Mycena* and *Hygrophorus*. *Lentinus* species with funnel-shaped 'mushrooms' show a preference for decaying timber and are often seen in large numbers on fallen trees.

Besides the saprophytic species, Basidiomycetes also form symbiotic mycorrhizal associations with forest trees. These mycorrhizal fungi include species of *Amanita* and *Boletus*. *Amanita* species are usually coloured and attractive, but they can be deadly poisonous.

Other Important Species

Prominent and attractive fruit bodies are also formed by the Ascomycetes, commonly known as cup fungi. Brightly-coloured species, unlikely to be missed in tropical rain forests, are *Peziza* and *Cookeina*. These fungi grow among the forest floor litter and form attractive beige, orange or pink fruit bodies.

PLANT PATHOGENIC FUNGI

Pathogenic fungi parasitise living plants and encourage the spread of disease of the infected plants. This results in economic losses, especially when the disease is of an epidemic scale. Serious crop diseases reported in Indonesia include 'blast' of rice caused by *Pyricularia oryzae*, and 'pokah bung' of sugar cane caused by *Fusarium moniliforme*. Other well-known plant pathogenic fungi include *Phytophthora* species which attack papaya, rubber and oil palm trees, *Pythium aphanidermatum* which causes 'damping off' of cabbage seedlings and *Albugo ipomoeae–aquaticae* (below) which attacks vegetables such as water spinach. The latter fungus appears as white masses of spores on the surface of the infected leaves, hence the name 'white rust' for the disease. Rain falling on the leaves facilitates the spread of the disease.

Members of three other classes of fungi — Oomycetes, Zygomycetes and Deuteromycetes — do not form large fruit bodies. These species are less familiar to most people, although they are economically important. The Oomycetes, in particular, include members which parasitise living plants and cause diseases of vegetables and crops. The Zygomycetes are commonly represented by *Mucor*, the nuisance bread mould. An important family of soil Zygomycetes is the Endogonaceae, members of which form endomycorrhizal associations with many plant species. Fungi not known to reproduce sexually are placed in the largest class of fungi, the Deuteromycetes. The main mode of reproduction for these fungi is by the formation of asexually produced microspores.

Fungi and Humans

While decomposition may be an ecologically important process, it can be a nuisance in everyday life. Nevertheless, people have capitalised on fungal strengths, using them to ferment raw materials and produce food, beverages, drugs, vitamins, and other useful chemicals. Also of interest are the edible mushrooms which can be cultivated under tropical conditions. These include padi straw mushrooms (*Volvariella volvacea*) and oyster mushrooms (*Pleurotus sajor–caju*).

Various fungi are used in the production of traditional fermented foods and beverages such as *kecap*, *tempe*, *oncom*, *tape*, *dadih*, *brem* and *brem bali*. *Kecap* fermentation uses a mixture of *Aspergillus oryzae* (Deuteromycete) or *Rhizopus oligosporus* (Zygomycete) and yeasts (Ascomycete). The fermentation of soyabeans to produce *tempe* uses species of *Rhizopus*, while fermentation of peanut cake to produce *oncom* is carried out with species of *Neurospora*.

Many fungi are sensitive to environmental changes. Major modifications of the environment, such as forest clearing, can result in the extinction of fungi. Little attention has been given to fungi in discussions on conservation and there are only a few specialists in Indonesia. Indonesian species are also poorly documented, but it is hoped that interest will increase before it is too late for many species.

A *Fungi growing on the remnants of a fallen tree.*
B Entoloma sp., *with its striking blue colouration.*
C Peziza sp., *an attractive cup fungus.*
D Lepiota sp., *common white mushrooms with brown scales.*
E *Tooth or Hedgehog fungi (*Hydnum repandum*).*
F *Bracket fungus (*Ganoderma sp.*).*
G *Hairy cup fungus,* Cookeina tricholoma.
H *Stinkhorn fungus,* Dictyophora indusiata.

MYCORRHIZAL FUNGI

Many fungi form mycorrhizal associations with the roots of plants. The fungi feed by secreting digestive enzymes from the network of slender filaments which takes the place of roots, and absorbs the soluble products of digestion. The mycorrhizal fungus enhances nutrient uptake by the roots from decomposing leaf litter, while deriving its nutrients from the photosynthates of the host plant. This symbiotic relationship is essential for the establishment and growth of seedlings, especially in nutrient–poor soils. Mycorrhizal infection of roots can be confined to the root surface (ectomycorrhiza) or located within the root cortical cells (endomycorrhiza) as illustrated at right. Mycorrhizal fungi have direct applications in agriculture and forestry.

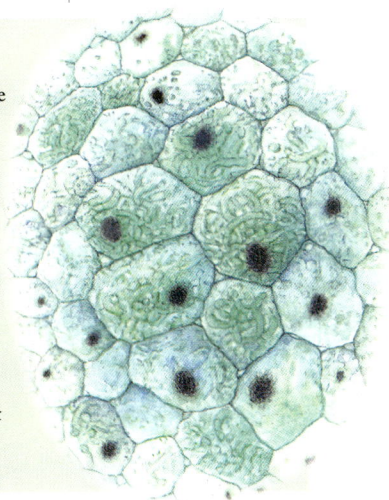

Lichens

Lichens are an ancient group of spore–producing organisms which are both widespread and easily recognised, though they are often mistaken for mosses. Part of the fascination of lichens lies in the fact that they are composite organisms, consisting of a fungal partner, and one or more photosynthetic or plant partners, growing together in a controlled and very specific way. This biological partnership of two dissimilar organisms is known as symbiosis.

Bright green filamentous lichens, which appear like cotton wool, are commonly found on the bark of forest trees and shrubs.

Distribution
Lichens enjoy a worldwide distribution. They are common in Indonesia, with over 1,000 species currently known and a likely, eventual total of 2,500. Globally, some 15,000 different lichens have been identified, but their diversity in tropical regions is still largely undocumented with numbers probably closer to 20,000 as more are collected and classified.

Active Partnerships
The lichen symbiosis is one of the most successful partnerships known in nature. The dominant partner is always the fungus, or mycobiont. The plant component, or photobiont, however, plays a vital role in the relationship in that it provides the fungus with food from photosynthesis, receiving in return a relatively stable environment in which to grow. This strategy has enabled lichens to colonise some of the most inhospitable habitats in the world, including polar and alpine regions. Lichens are found in almost every conceivable habitat ranging from the seashore to hot and cold deserts, forests, grasslands, swamps and even on manmade substrates.

Varied Growth Forms
Lichens exhibit several growth forms identified as crustose, foliose, filamentous, and fruticose. Crustose are closely attached to the substrate, often being immersed in it. They form distinct mosaics on rocks, walls and on the bark of tropical trees. Foliose lichens are differentiated into an upper and lower surface. The latter attaches itself to the substrate, while the former is the site of photosynthesis when the lichen is wet. Foliose kinds grow up to 15 to 40 centimetres in diameter in favourable conditions. Filamentous lichens look like cotton wool, grow on the bark of trees and shrubs and are orange–yellow or bright green in colour. Fruticose forms attach to the substrate by one or more root-like holdfasts and may be either pendulous or upright. They have distinctive chemical mechanisms for dealing with high light intensities and are found in the tree canopy. Other chemical compounds act as a defence against insect or fungal attack; several of these have vital antibiotic or anti–cancer properties.

A Structured Growth
Lichens are organised into various layers consisting of an upper cortex of closely compacted fungal cells, overlying a photosynthetic layer containing green algae and/or blue–green algae. Beneath the photosynthetic layer lies the medulla, an often extensive tissue of loosely arranged fungal threads which are important for water storage. Foliose lichens have a lower cortex, which is similar in structure to the upper cortex but which also produces specialised cells that anchor the plant body (thallus) to the substrate. Crustose and filamentous lichens have no lower cortex.

An example of lichens with the crustose growth form which are closely attached to, or immersed in, the substrate. As above; they are recognised as distinct mosaics on rocks, walls or bark of tropical trees.

Lichens reproduce often by asexual means: Phyllidia, are more organised miniature lichen–like propagules found on the margins and upper surfaces of the lichen, Sticta.

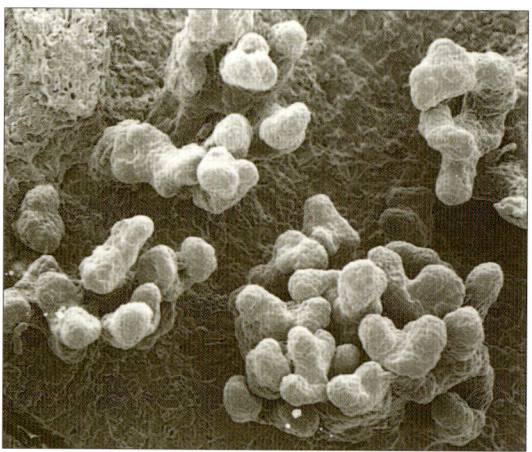

LICHENS AS NATURE'S FERTILISERS
Many tropical lichens have a blue–green algae (cyanobacterium) as their photosynthetic partner. This gives the lichen a distinctive, dark blue–grey to blackish, or greenish–black colour. The genera *Coccocarpia* and *Lobaria* (as below), which are widespread in both tropical and temperate forests, are examples of cyanobacterial lichens. Cyanobacterial lichens differ from other lichen species which have a green algae photobiont, in that they are able to grow in low light intensities and, more importantly, they are able to "fix" atmospheric nitrogen. In this way they can transform inert, gaseous nitrogen from the atmosphere into organic nitrogen–rich compounds, including amino acids, the precursors of proteins. By contributing to the circulation of nitrogen through the ecosystem, cyanobacterial lichens fulfil a fundamental role in the ecology of rain forests; a role that has only recently been recognised for these organisms.

DIVERSITY OF INDONESIAN FLORA

Reproduction Patterns

Lichens, unlike flowering plants, reproduce mainly by asexual means. This may result from the thallus breaking up, but more commonly involves the formation of vegetative propagules containing both photobiont and mycobiont cells. These propagules may be relatively heterogeneous, appearing as flour-like granules (soredia) on the upper surface or at the margins of the parent organism. Alternatively they may involve more complex structures that take the form of miniature lichens (isidia or phyllidia) which again appear on the margins and upper surfaces of the parent lichen. In many instances spores are distributed by specialised fruiting bodies (apothecia or perithecia). These subsequently germinate and form new lichens when a suitable photobiont is secured.

Environmental Indicators

Lichens are very sensitive to external environmental changes and, in this respect, can serve as monitors of atmospheric and terrestrial pollution, acid rain, heavy metals, radioactive fallout and ultraviolet radiation as a result of decreasing ozone levels. Lichens are also used to date rock surfaces in alpine areas and can be used to monitor environmental changes due to global warming.

The success of lichens as environmental indicators is closely connected with the delicate symbiotic balance existing between the dominant fungal partner and the photosynthetic partner which provides the lichen with its food. Any environmental change which upsets the vitality or viability of the lichen photobiont, with most atmospheric pollutants being detrimental to it, leads to the withering and death of the whole lichen. Since pollution–sensitive lichens respond to levels lower than those producing a response in trees or other higher plants, they can be used as effective monitors of pollution produced from cars or from industrial or domestic smoke. In many instances lichens produce photoprotective chemicals in their upper cortex to protect the photobiont layer from radiation damage. In forest canopy species in particular, these photoprotective pigments are very important since many species produce them in response to levels of ultraviolet radiation.

Work done on lichens in Antarctica has direct relevance to their persistence in other biomes including temperate and tropical rainforests. If lichens are absent from an area where they once were found, it indicates that the area has been subjected to high pollution loading and other components of the biota of the area can also be at risk. Areas of high pollution density are often referred to as 'lichen deserts'.

STRUCTURAL COMPOSITION OF LICHENS

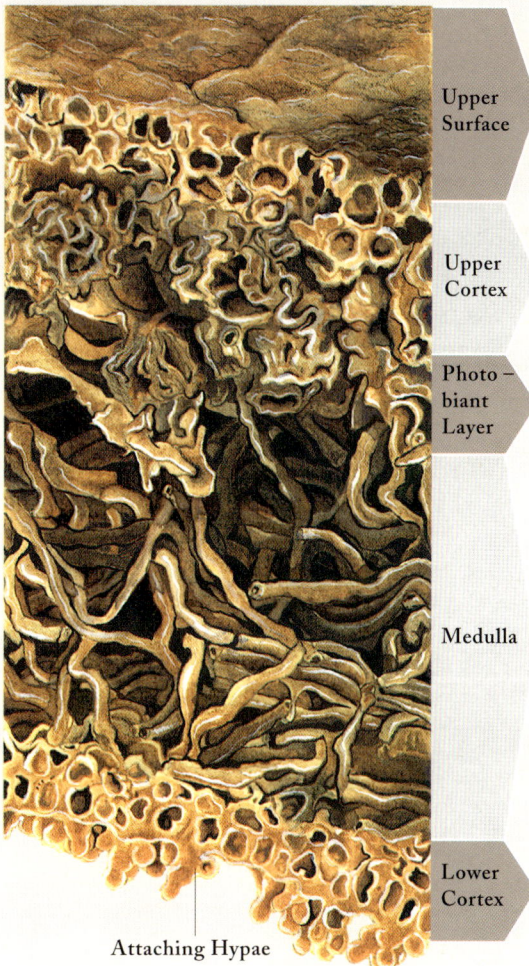

Attaching Hypae

The lichen symbiosis is one of the most successful known in nature. It consists of the mycobiont (fungus) and a photobiont (green algae or cyanobacteria).

The structure of the lichens is stratified into the various components (left): The upper surface where the vegetative propagules are found. The dominant partner which provides a stable environment for growth is found as a closely compacted layer as the upper cortex. The photobiont layer houses the green algae and/or cyanobacteria which provides food from photosynthesis. The medulla is an extensive tissue of loosely arranged fungal threads involved in water storage. The lower cortex, absent in crustose and filamentous lichens is present in foliose lichens for anchorage.

Lichen–draped tree in montane forest, 2,000 metres above sea level, in the Baliem Valley region of Irian Jaya.

LICHENS ON LEAVES

In the tropics it is common to see lichens growing on the living leaves of a variety of trees and shrubs. These foliicolous lichens, as they are termed, are very fast growing in contrast to lichens found in most other habitats which tend to persist for long periods of time. The rather short life cycles of foliicolous lichens growing on short–lived substrates like leaves (right), makes them useful in the study of the establishment and development of lichen communities. Foliicolous lichens often grow on leaves rich in aromatic compounds, so that these may have a potential use as indicators of biologically active compounds in leaves.

Bryophytes

The term 'bryophytes' loosely refers to a group of non–flowering, green plants, characterised by little or no organised vascular tissue that share a common and unique life cycle. The group comprises mosses, liverworts and hornworts of which there are at least 1,500 species represented in the Indonesian flora. Bryophytes vary greatly in form, size and colour. In montane areas, where they are particularly abundant, they may be the most colourful element of the canopy forest flora.

The pioneer genus, Marchantia, a liverwort common to earth banks produces conspicuous structures containing the asexual propagules — gemmae which are little bundles of cells.

Meteorium sp., a very fern–like and pendulous moss found in South Sulawesi.

Beauty in Miniature

The majority of bryophytes are too small to be seen by the unaided eye, but a simple hand lens with ten times magnification opens up a whole new world of remarkable diversity and beauty. A compound microscope is required for the identification of all the different species, but is unnecessary in order to gain appreciation for the range of species and basic lifestyles of many bryophytes. In appearance, mosses and liverworts are all small, green plants usually no more than a few centimetres long. These plants consist of either a leafy stem or a flat, ribbon–like body called a thallus. They are anchored to the soil by long, thread–like rhizoids instead of true roots.

Asexual Reproduction

Being relatively simple in structure, mosses can readily reproduce vegetatively from broken off fragments: wind disturbance or a bird collecting nesting materials are two of the many agencies which can transport pieces of bryophytes over moderate distances. Another form of vegetative reproduction displayed by bryophytes is the production of special reproductive bodies known as gemmae. Gemmae are unicellular or multicellular bodies, often budded off from specialised generative tissues on stems, leaves, rhizoids or protonemata. This is the first stage in the development of the gametophyte. Gemmae provide a rapid and effective means of short–range dispersal.

Sexual Reproduction

A disadvantage of vegetative, as opposed to sexual, reproduction is that because the offspring are identical to the parent, adaptive genetic evolution to changing

THE LIFE CYCLE OF A MOSS

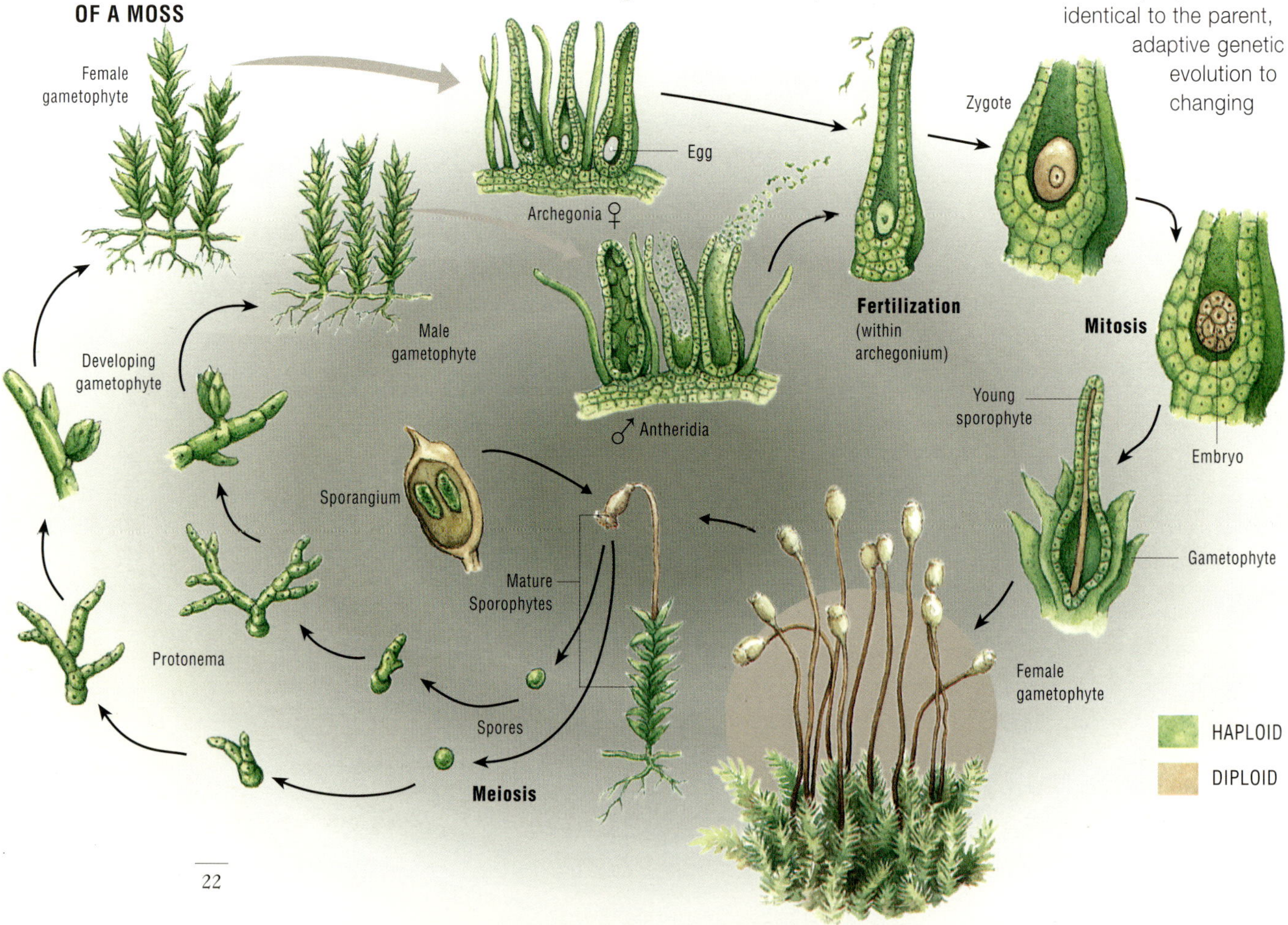

circumstances of climate or habitat is precluded. To help overcome such drawbacks, bryophytes may also reproduce sexually. This process, known as an 'alteration of generations' includes two generations of plants: a sporophyte phase which produces spores and a gametophyte phase, which produces male and female sex cells (gametes). In bryophytes, the gametophyte generation, which developes from asexual spores, is dominant, while the sporophyte generation is reduced. In this way, the bryophytes resemble simple green algae.

Liverworts

Liverworts vary greatly in size but can usually be recognised by their leaves: those of leafy liverworts, for example, have a midrib and are always arranged in two or three rows with the cells of the lower layer differing in structure from the remainder. The bodies of some liverworts are divided into lobes. Thalloid liverworts, in contrast, lack true leaves, and are otherwise undifferentiated in terms of specialised structures or tissue systems. They are usually flat, green plants which may be either simple or branched.

Hornworts

Hornworts are leafless plants found commonly on moist soil along riverbanks and streams. The gametophyte generation is typically flattened, irregularly lobed structure, with the sexual organs sunk into the upper surface of the thallus. Root-like structures on the under surface of the plant anchor it to the substrate. Hornwort sporophytes are perennial structures that grow continually from the base, splitting above to continually release spores. They are dependent on the attached gametophyte for water and nutrients.

Montane Adaptations

A characteristic of many mosses is their ability to regulate their metabolic processes in response to moisture levels in the environment. In situations of extreme dehydration they are able to simply switch off altogether. In a desiccated and dormant state, a moss is able to withstand high temperatures and may survive for considerable periods.

At high altitude, cool temperatures and leached soils slow down the growth of trees, but encourage that of bryophytes. In the cloud forest above 3,000 metres, mosses may constitute more than half of the living biomass. A dense moss carpet holds many times its own weight and volume of water, and regulates the through-flow of water within the ecosystem.

In the humid environment of montane forests, many bryophytes assume different and larger, more

spectacular growth forms designed to catch as much light as possible. Many unrelated groups have to produce similar forms in response to the same habitat parameters — a case of convergent evolution.

Playing a Vital Role

Bryophytes are able to colonise tree bark, rocks and other generally inhospitable surfaces. In doing so they provide a crucial habitat for a huge variety of other plants, as well as vast populations of invertebrates. The survival of many orchid species, for example, would not be possible in the absence of a healthy bryophyte cover. Even evergreen leaf surfaces may be exploited by flat-growing epiphyllous liverworts, most of which belong to the family Lejeuneaceae. Being small and delicate, such plants are short-lived. Their nutritional requirements are minimal and are probably supplied mainly by minerals eluted from canopy foliage.

The protective environment of tropical rain forests has allowed the survival of relatively unadapted bryophytes which have maintained many primitive features. The 'living fossil' *Treubia* shows very little change from fossil forbears dating back to the Carboniferous Period, some 3,000 million years ago. Despite such constancy, these types of plants would disappear very rapidly if forests were felled or their environment otherwise damaged.

Spiridens (left), a spectacular moss with a 'shuttlecock' form commonly found on the trunks of tree ferns.

Anthoceros sp. (right) demonstrating distinct features of the hornwort sporophyte — an elongated horn-like capsules that grow from the mat-like gametophyte.

Calymperes sp. in its natural habitat with detail of a leaf (far left): gemma of some species are very sticky and are dispersed by insects, especially ants.

A view of a moss forest in Irian Jaya: a healthy bryophyte cover is vital to the survival of many orchid species, and also supports large populations of invertebrates.

Ferns

Ferns and allied plants, such as clubmosses, horsetails and quillworts, are traditionally grouped together as pteridophytes. They are descended from some of the earliest known plants which appear in fossil records some 400 million years ago. Almost all are terrestrial species, growing on soil, rocks, or even other plants; a few are aquatic. Water is, however, an essential prerequisite for sexual reproduction in all species.

Clubmosses demonstrating features such as small leaves, bearing one sporangium each, are common as pioneers in bare or disturbed sites (from top): Selaginella *sp. (Selaginellaceae).* Lycopodiella cernua *(Lycopodiaceae).*

Equisetum ramosissimum *subsp.* debile, *the only horsetail found in the Malesian region.*

»»*The spectacular crown of the tree fern (*Cyathea latebrosa*) supported on a tall trunk. Tree ferns are found in open forests in the lowlands and hills.*

Early Plants

The ancient evolutionary history of ferns is reflected in the fact that they are the only group of vascular plants whose reproductive cycle still includes both an asexual, sporophyte phase and a sexual, gametophyte phase. Because ferns, clubmosses, quillworts and horsetails all share these very generalised characteristics they have traditionally been grouped together. Current thought, however, concludes that they are only distantly related to one another, and not through a relatively recent common ancestor as was once supposed.

Structure and Life Cycle

Ferns are generally small plants, growing perhaps to a height of two metres, although some of the larger tree ferns may reach 25 metres. Most species comprise a rooting system, rhizomes and leaves, known as fronds, which grow as their coiled tips unfurl. Ferns possess a vascular system which is less developed than that of flowering plants.

Ferns and allied plants exhibit a complicated life cycle that involves the development of two distinct plants which bear no resemblance to each other, and exist independent of one another, and of two phases of reproduction — a gamete-producing stage (gametophyte) and a spore-producing phase (sporophyte). The gametophyte stage, which is less than one centimetre long and usually short-lived, represents the sexual generation of the fern. The sporophyte phase is easily recognisable as a fern and bears fronds which asexually produce spores for dispersal and colonisation of other sites; the sporophyte may live for decades.

Major Groups

The Malesian region has a complex geological and biogeographical history which has led to the evolution of a number of endemic genera of flowering plants. Perhaps, because of their earlier origin, ferns in this region have more endemic species than genera. Delimitation at the family level of classification is still not clear for most of them.

LIFE CYCLE OF A FERN
The life cycle of a fern and allied plants involves two distinct and independent plants: the gametophyte and the sporophyte which are involved in sexual and asexual reproduction respectively as illustrated above.

It is estimated that there are about 4,000 species of pteridophytes in Malesia, the great majority of which are ferns. These generally have large leaves, each bearing many sporangia — organs that produce asexual spores. In contrast, the leaves of clubmosses (Lycopodiaceae and Selaginellaceae) and quillworts (Isoetaceae) are small and bear only one sporangium each. Clubmosses are often pioneer plants, colonising bare or disturbed ground. The quillwort (*Isoetes*) has a very different habitat in that it grows as a submerged plant in high altitude lakes.

Some families, however, are easily recognised. The filmy ferns, Hymenophyllaceae, grow in habitats with high humidity. This is essential because the leaves of these ferns lack a protective cuticular layer.

DIVERSITY OF INDONESIAN FLORA

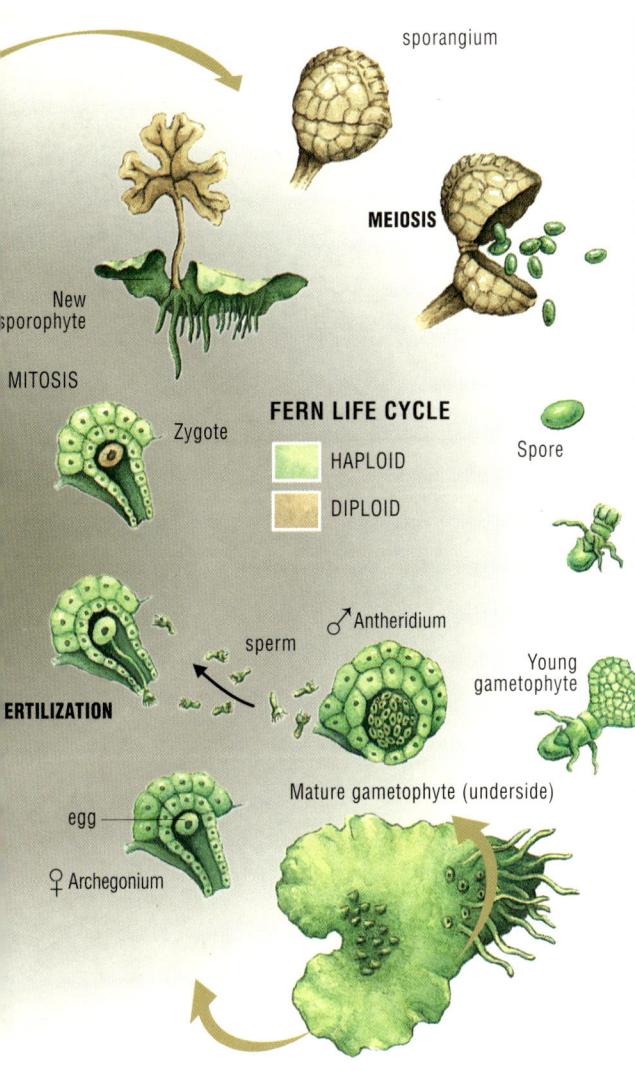

FERN LIFE CYCLE
HAPLOID
DIPLOID

TREE–TOP GARDENS

Some species of ferns grow as spectacular plants on the trunk and branches of trees at a height of up to 30 metres above the ground. They depend on light, rain and probably, exudates from other plants for nutrients; but are not parasites on the trees. A single bird's-nest fern (*Asplenium nidus*), right, may weigh over 100 kilograms, so it is not surprising to find such ferns lying on a forest path because they have overbalanced or the supporting branch has broken off the tree.

Their fronds (leaves) either have expanded bases (as in *Asplenium*) or are completely modified into specialised 'shield' fronds, as found in the well–known stag's horn fern (*Platycerium*), which trap falling leaves of the forest trees which then decay to form humus. This effectively produces a 'plantpot in the sky', other ferns, clubmosses, orchids and screwpines can grow here. Ants and earthworms are found in these aerial gardens.

Such delicate leaves react very quickly to local changes in the environment; cutting a new trail through the forest can cause adequate disturbance to kill tiny, filmy ferns growing on tree boles, banks of roads or boulders. In contrast, species of Gleicheniaceae seem to thrive in more inhospitable habitats, and quickly colonise natural landslips or the banks of new roads: even in the sulphur–laden environment of a volcanic crater, typical species of *Gleichenia* are located.

Members of the Thelypteridaceae and Dennstaedtiaceae families are often counted as weeds. They also quickly colonise disturbed areas and abandoned kitchen gardens or plantations.

Tree ferns *(Cyatheaceae)* are perhaps the most spectacular and one of the most easily recognised groups of ferns. They grow in lowland forest communities, but are more striking at higher altitudes, where the land is often subjected to frequent fires through natural or man–made causes. Here they may be the major 'tree' component of the local vegetation. The young, tightly furled fronds are protected against fire by their coating of dense scales and by the older unfurled and expanded fronds. The outer zone of a tree fern 'trunk' is a mass of roots that carry water and nutrients from the ground to the crown of fronds. These trunks have proved an ideal medium for growing orchids.

Unfortunately the demand for this material by commercial orchid growers has been so great that all tree ferns are now protected by an international convention regulating trade in endangered species — The Convention on International Trade in Endangered Species of Wild Fauna and Flora (CITES).

Ferns are highly versatile and are able to colonise a wide range of environments (from top):

Filmy ferns, a species of Hymenophyllum, *that can only survive in moist conditions such as those found in the lower montane forest. This is due to the absence of a protective cuticular layer.*

Gleichenia, a species of Gleicheniaceae, is able to thrive easily in harsh environments. These include natural landslides, banks of new roads, and even the sulphur–laden surroundings of a volcanic crater.

Plants Which Bear Seeds

The vegetation of most of the world's land surface is dominated by seed-bearing plants, with perhaps 25,000 species occurring in the Malesian region alone. Seed plants differ from other vegetable forms in that their eggs are fertilised while they are still attached to the mother plant. The developing embryo is subsequently retained until it is ready for dispersal. This process appears to have been favoured through evolution as seed-bearing plants are now so common and widespread.

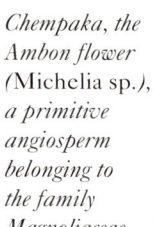

*Chempaka, the Ambon flower (*Michelia sp.*), a primitive angiosperm belonging to the family Magnoliaceae.*

Double Fertilisation

The majority of seed plants are angiosperms or 'flowering plants', whose seeds are almost always enclosed in a box-like ovary on top of which are a style and stigma. Pollen tubes germinating from male pollen grains deposited on the stigma by wind or animals, grow down through the style towards the ovary. These tubes convey the male nucleus of the pollen grain to the female nucleus of the ovary and, in the process, probably compete with one another for access to the egg sac. It is possible that the tissue of the mother plant plays a role in the selection process, thereby ensuring an optimal match for her eggs. A second nucleus from the successful pollen tube unites with two non-egg nuclei in the egg sac, giving rise to the endosperm, a nutritive tissue which provides food for the developing embryo. In this respect, the reproductive process of flowering plants comprises a 'double fertilisation'. That the endosperm arises thus, is perhaps the only feature which clearly demarcates the 'flowering plants' from other seed plants.

Angiosperms, the only plants found from land to sea, clockwise: Rhizophora *sapling,* Dysoxylum arborescens, Rhodomyrtus tomentosa *and rice (*Oryza sativa*).*

Enormous Variety of 'Flowering Plants'

Flowering plants range from the tallest trees known, for example the eucalypts, to some of the smallest multicellular ones such as duckweed. The structure of the flower varies considerably between species. Attached to the end of a stalk, flowers often bear distinct, usually green, sepals which protect the more delicate, often brightly-coloured and patterned petals which are attractive to insects and small birds. Male and female reproductive organs are arranged inside the flower head: pollen grains are produced from male stamens which surround the female part of the flower — the carpel — which contains the eggs. The latter, once fertilised, develop into embryos, with the surrounding tissues becoming the seed coat and the carpel walls enlarge to accommodate the swelling embryos. There are, however, many exceptions to these generalities, particularly in wind-pollinated species where the flowers may be very small and green, and the petals greatly reduced.

Angiosperms are usually divided into two sub-classes based on the number of seed leaves (cotyledons) they produce. In general, monocotyledons have a single seed-leaf with parallel veins, and flowers whose constituent parts come in threes, or multiples thereof. Dicotyledons, on the other hand, possess two seed leaves with net-like venation, and flower parts in five. Again, there are many exceptions to this rule. Peppers, for example, have a single seed-leaf, yet are classified as dicotyledons, while yams, which have leaves with net-like veins, are placed with the monocotyledons. The latter group includes palms, pandans, grasses, sedges and orchids. The much larger grouping of dicotyledons includes the Dipterocarpaceae (dipterocarp trees), Compositae (daisies), Moraceae (figs), Leguminosae (beans) and Euphorbiaceae (spurges).

Virtually all familiar plants of the Indonesian landscape are angiosperms. They include the trees of the rain forest, palms, fruit trees, rice and other grasses, vegetables and orchids. Together they provide almost all of the fibres, timbers, vegetable foods and beverages used by people, whose evolution and history have always been inextricably linked with the plant world and its products. No major civilisation has arisen without the exploitation of starch-rich grass seeds: rice in Southeast Asia, maize in the Americas, wheat and other grains in the Near East regions.

Angiosperms range in size from minute, rootless aquatic plants to emergent rain forest trees, and climbers that may grow up to 150 metres in length in the case of rattan palms. Compared with other seed plants, angiosperms are remarkable for the diversity of their life forms which include herbs, mycotrophs (plants living symbiotically with a fungus) and parasites — habits which are virtually unknown among other seed plants. Angiosperms have even re-invaded the sea, both as herbaceous plants — seagrasses grow to depths of 30 metres — and as trees in mangrove swamps. They are also notable for their elaborate interactions with animal species, which may act as either pollinators and/or dispersal agents.

PRIMITIVE ANGIOSPERMS

Certain features of angiosperm flowers are considered archaic. These include indefinite numbers of spirally arranged parts in the flower; large seeds with brightly-coloured aril coats; and homogenous woody tissue which is not differentiated into strengthening cells (fibres) and water-conducting cells. Indonesian families with large numbers of these, and other, supposedly primitive features include the Winteraceae, Magnoliaceae, Myristicaceae and Nymphaeaceae.

ADVANCED ANGIOSPERMS

Angiosperms possessing several characteristics that are considered to have evolved relatively recently are said to be 'advanced'. One such characteristic is a reduction in the number of floral parts which are often united to create a floral tube; these plants are most often visited by long-tongued insects. Other advanced features include sophisticated defence chemicals and small seeds with a seed coat consisting of a tube (pachychalaza) which encloses the developing ovule. The latter characteristic is an entirely new feature and totally unrelated to the seed coat of more primitive plants. These advanced angiosperms are very successful globally, particularly as herbaceous plants on open ground. Many are regarded as weeds and are toxic to livestock. The largest angiosperm family is the Compositae, an advanced group whose members are characterised by heads of tiny tubular flowers superficially resembling one simple flower. An example of an advanced angiosperm, *Acanthus ilicifolius* (right) is a bird pollinated prickly shrub of the mangroves.

Gymnosperms

All other seed plants are grouped into a category referred to as gymnosperms, although many of these are probably not closely related. The common characteristic of gymnosperms is that they bear naked seeds which are not enclosed by an ovary. Many gymnosperms, however, show features which are thought of as being typical of 'flowering plants' and, in this respect, it is possible that some should be considered as part of the same lineage. This applies in particular to members of the Gnetopsida, whose leaves and wood structure are often confused with angiosperms. Gnetopsida are represented in Indonesia by the genus *Gnetum*, which is largely made up of woody climbers or lianas.

The best known sub-class of gymnosperms is the Pinopsida or conifers. Conifer leaves are either needle- or scale-like and their reproductive organs are in the form of cones. Male cones are covered with fertile scales, each containing pollen sacs, while female cones consist of a sterile, outer scale, or bract, and an inner ovule-bearing scale. Wind-borne pollen is received by the open scales of the female cone, which then close, trapping the pollen within; actual fertilisation may not take place until several months later. Upon reaching maturity the woody female cone may either open on its own accord to release its seed, or else remain sealed until opened by some external agency such as foraging animals or fire. In some species, the scale bearing the nut-like seed may act as a wing to assist airborne dispersal.

Exotic pine species, which are grown in plantations throughout Indonesia, are perhaps the most familiar gymnosperms in the region. There are, however, numerous indigenous species such as *Araucaria* and *Agathis* spp., as well as native pines and podocarps. Their wood differs from most angiosperm wood in having only one cell, as opposed to living fibres and vascular systems, which act as support and transport systems respectively.

Cycads and Ginkgos

The two other extant groups of gymnosperms — the Cycadopsida (cycads) and Ginkgoopsida (gingkos) — are a little less familiar. Cycads are stout, slow-growing trees, usually with unbranched stems and pinnate leaves. Their roots often harbour blue-green algae. Their reproductive organs are in the cones (except for female *Cycas*), with male and female cones occurring on different trees. These plants are pollinated by beetles and/or the wind. Cycads have large seeds and the seedlings have two cotyledons. Cycads are much sought after for horticulture and have become endangered through over-collecting. Some species of cycad yield a form of sago from the trunk.

The deciduous maidenhair tree (*Ginkgo biloba*) is the only living member of Ginkgoopsida. Its fan-shaped leaves are similar to fossils which date from 200 million years ago. This species does not grow naturally in Indonesia, but its dried seeds are imported. These can be roasted and eaten; they are also a source of oil and insecticide.

The fallen seed of Ginkgo biloba. *Fossils of this species have been found that are over 200 million years old.*

Male and female Cycas *(top to bottom): the dumpy trees have terminal cones in the males, and lateral seed-bearing organs in the females. Although sago is obtained from the trunks of several species, the seeds are known to contain carcinogenic substances.*

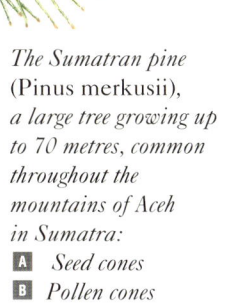

The Sumatran pine (Pinus merkusii), *a large tree growing up to 70 metres, common throughout the mountains of Aceh in Sumatra:*
A *Seed cones*
B *Pollen cones*

The World's Largest Flowers

Indonesia's rain forests, one of the oldest and richest on earth are dominated by dipterocarp trees in whose shelter have evolved plants with giant flowers that rank among the greatest wonders of the natural world. Two plants in particular are renowned not only for their size, but also for their spectacular shapes, textures and odours — Rafflesia arnoldii and Amorphophallus titanum.

Titan arum, Amorphophallus titanum *featured on the Indonesian 500 Rupiah bank note.*

The blooming phases of Rafflesia arnoldii *from the bud to mature flower ❶ to ❹ — The bud appears like a small orange–brown cabbage which gradually swells and eventually bursts open as a dark red bloom. The blooming process takes about 24 hours. The five leathery petals open one by one around a central domed cavity. Flies are attracted by its foetid smell, and guided by light 'windows' in the diaphragm.*

Giants of the Plant World

These plants were brought to the notice of the world in the 19th century. *Rafflesia arnoldii* was named after Sir Stamford Raffles, founder of Singapore, and his physician Joseph Arnold, who found it while exploring the foothills near Benkulu in 1818. An Italian explorer, Odoardo Beccari, was the first to describe the titan arum, *A. titanum*, in 1878. In both instances, their descriptions were initially greeted with disbelief. *Rafflesia* flowers are one metre across and weigh nearly seven kilogrammes, while the titan arum's inflorescence grows to over two metres in height (3.3 metres has been recorded). Over 100 species of *Amorphophallus* grow in various parts of the world, many of which have large, grotesque blooms. *A. decus–silvae,* also found in Sumatra, has a smaller inflorescence than *A. titanum,* measuring 1.5 metres, but on a much longer stalk, giving an overall height of some 4.4 metres.

In the case of *R. arnoldii* the flowers are produced singly, and are either male or female. *Amorphophallus*, on the other hand, produces a spike–like inflorescence on which many flowers are clustered. The upper section, or appendix, is sterile, but below this, hidden from view, are separate bands of numerous tiny male and female flowers.

The Complete Parasite

R. arnoldii is a parasite of the woody liana *Tetrastigma*, which loops and twists for great lengths through trees in the rain forest. There are 15 or so other species of *rafflesia,* all of which are found in Southeast Asia, although three are believed to be extinct. Unless flowering or fruiting, Rafflesia is invisible, consisting of microscopic filaments which grow inside the host liana, absorbing nutrients from the liana's sap. The minute seeds probably gain entry to the vine through damaged bark or roots, which have perhaps been trodden upon or clawed by an animal. Several years elapse before flowering begins. At this point, tiny orange–brown, cabbage–like buds erupt through the bark and grow to a huge size.

The open flowers last only a few days. Their five leathery, warty lobes, surround a domed cavity which has a capacity of some 6.5 litres. A central column is covered by a disk of spines, the function of which is not known. Flies are attracted to its foetid smell and guided by light 'windows' in the diaphragm. If a female flower is pollinated, a spongy fruit develops which takes six months to ripen. It is about 15 centimetres wide and contains thousands of seeds which are roughly a millimetre in length. Their means of dispersal remains a mystery — possibly they are spread by termites or else trampled to another site by animals that dig for food.

The Corpse Flower

Amorphophallus grows from a corm, which can reach 50 centimetres in diameter and weigh over 50 kilogrammes (the heaviest recorded was 75 kilo–grammes). After a dormant period of several months, the bud develops at an astonishing speed of four to 20 centimetres a day. When fully developed, the spathe — a bract that envelops the flower spike —

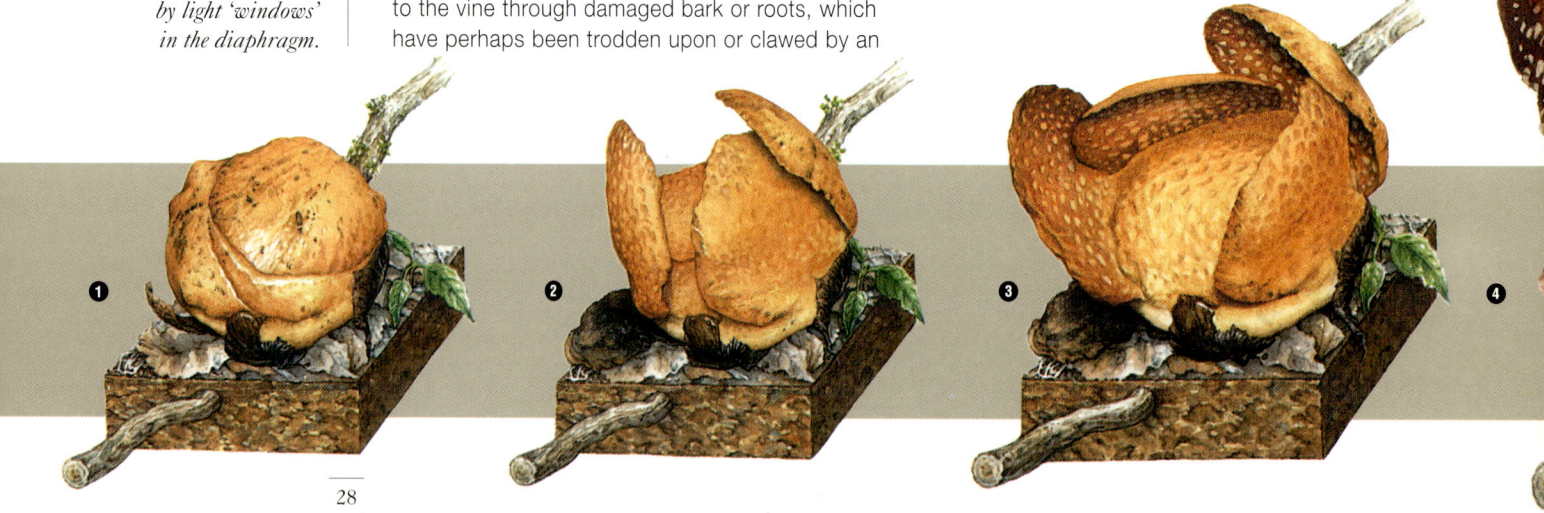

DIVERSITY OF INDONESIAN FLORA

opens like an upturned bell to expose the appendix.

Dung and carrion beetles home in on the stench wafting from the appendix and are attracted by the colour of the spathe, which resembles dried blood. Moving to the lower part of the flower they enter a chamber where, if carrying pollen from another bloom, they fertilise the female flowers. As the insects try to leave the chamber, they are coated with pollen released by the male flowers. Flowering lasts two to three days after which spadix and spathe collapse, allowing the insects to escape, and covering the chamber to keep the fertilised flowers dry.

As the fruit develops, the spathe and spadix decay, revealing a spike, about 50 centimetres long, of olive-sized fruits that turn red when ripe. The fruits are dispersed by birds, among them hornbills, and possibly mammals. After flowering, a solitary leaf is produced. It has a smooth, mottled stalk up to five metres tall and a trilobed, compound leaf about four metres across and 15 metres in circumference.

A Rare Event

Attempts have been made to cultivate rafflesia by inoculating vines in botanic gardens with seeds and by cutting sections of vines bearing rafflesia buds. While the former method has met with limited success, the latter is doomed to failure as the buds need a growing vine to survive. Damage to the vines reduces the wild population of rafflesia further.

A. titanum has been grown at several botanic gardens, mostly in containers from mature corms sent from Sumatra (export of corms and seeds is now illegal). Plants seldom live long because the corms tend to rot. The best record is held by the Bogor Botanic Gardens, West Java, where the species blooms every few years — an event which brings thousands of people flocking to the garden.

HEAT AND ODOUR IN ARUM LILIES

Amorphophallus plants, such as *A. paeoniifolius* from Java shown here, belong to the arum lily family or aroids (Araceae). An unusual characteristic of many arum lilies is that the appendix generates heat. Heat generation is triggered by a plant hormone and fuelled by reserves of starch in the massive corm. This causes chemicals to evaporate producing a foul odour. The *titan arum* produces one of the most powerful and disgusting odours of any plant. This is due to amines, including skatole (which gives a strong faecal smell), and ammonia. These compounds induce nausea and make the eyes water. The appendix attains a temperature of fur degrees Celsius above its surroundings, and the foul odour wafts through the still, humid air of the forest. This strategy has great survival value in that it attracts pollinators to a species with widely scattered individuals, which flowers infrequently and briefly, and requires cross-pollination.

Saving the Giants

R. arnoldii is found in Sumatra and Sarawak mostly in mixed dipterocarp forest on steep, moist hillsides and valleys below 1,300 metres. A single vine develops flowers at stages, although buds often drop off when they become too heavy. *A. titanum* is endemic to Sumatra, and grows in similar locations and in secondary forests from 120 to 365 metres. Sparse colonies are formed with individuals at various stages of growth. Flowering and fruiting are uncommon. Dormant plants may survive forest clearance, but are unlikely to thrive in brighter, drier conditions.

These endangered species are highly vulnerable to damage. Rafflesia is especially at risk from collectors of buds and flowers as specimens and for medicinal use. As the life cycle of both species is poorly understood, and cultivation near-impossible, protecting the forests is the only way to save these wondrous giants.

Rafflesia hasseltii, *located at an altitude of 450 metres in Upper Merangin, between Temiai and Muaraimat in west Jambi.*

COMPOSITE SECTION OF MALE AND FEMALE RAFFLESIA FLOWERS
1. Lobe
2. Diaphragm
3. Processes
4. Disk
5. Ramenta
6. Central column
7. Anther
8. Ovary
9. Cupule

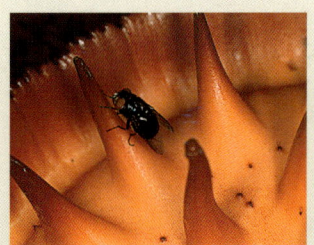

SURVIVAL TACTICS

The *Rafflesia* is protected by law and is designated the National Rare Flower (*Puspa Langka*) of Indonesia. It is pollinated by flies (top), attracted to the rotting meat-like smell. The tree shrew, seen investigating rafflesia (above), is probably one of the dispersal agents responsible for the continuation of this species.

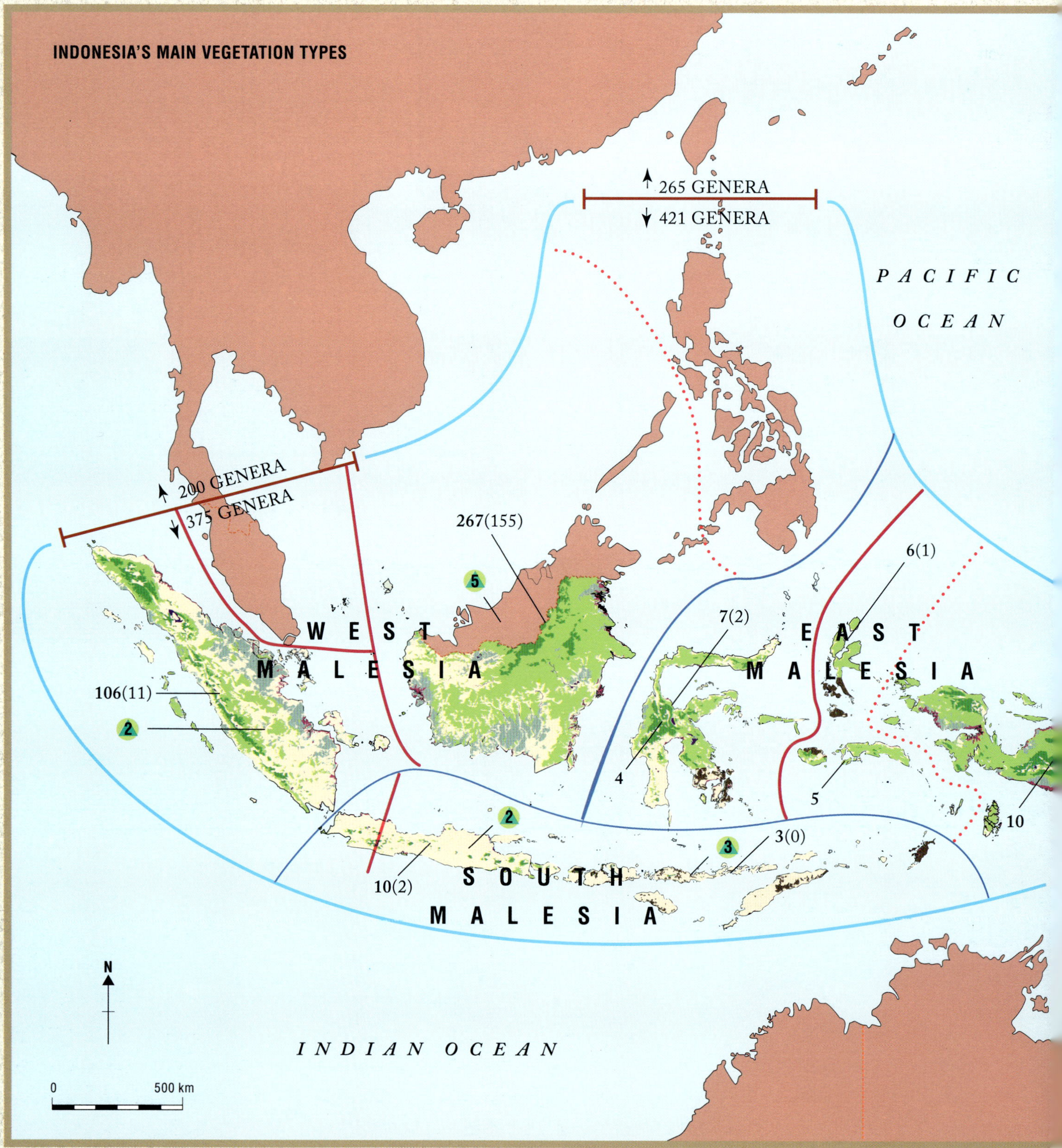
INDONESIA'S MAIN VEGETATION TYPES
265 GENERA
421 GENERA
PACIFIC OCEAN
200 GENERA
375 GENERA
267(155)
5
WEST MALESIA
106(11)
2
7(2)
EAST MALESIA
6(1)
4
5
10
2
3
3(0)
10(2)
SOUTH MALESIA
N
INDIAN OCEAN
0
500 km

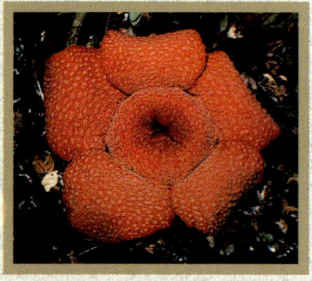

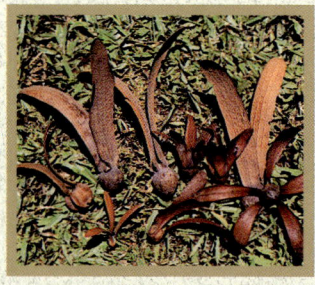

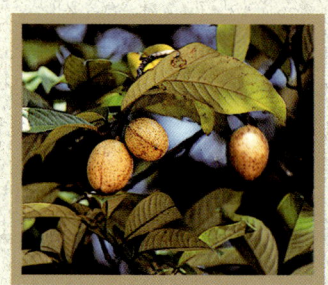

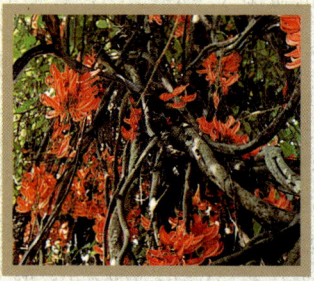

Legend

■	Water bodies
■	Mangrove
■	Inland swamp
■	Lowland rain forest
■	Montane seasonal forest
■	Lowland seasonal forest
■	Non–tropical forest
X	Number of dipterocarp species
(Y)	Number of endemic dipterocarp species
Z	Number of she–oak or casuarina species
—	Boundaries of Malesia
—	Subdivision boundaries of Malesia
···	Division boundaries of Malesia
⊢	Demarcation line

Representative flora from different parts of Indonesia (from left): The rafflesia flower (Rafflesia achensis) from Aceh in West Sumatra. Winged fruits of the dipterocarps, with a strong concentration in Western Indonesia especially in Kalimantan. Nutmegs (Myristica fragrans) from the Banda Islands in Maluku. The flame of Irian (Mucuna bennettii), a spectacular climber from Irian Jaya.

VEGETATION TYPES AND BIOGEOGRAPHY

Indonesia, together with Malaysia, Brunei, the Philippines and Papua New Guinea, form a coherent floral region called Malesia. The plants of this region are sharply differentiated from those of adjacent areas by narrow zones where there is a dramatic change in flora, and where many genera reach their distributional limits. Such transitions are also reflected in climatic changes, from permanently humid to more seasonal climates. Although there is no floral equivalent to Wallace's Line, there are some regional variations in the occurrence of certain species. Dipterocarps, for example, are found mainly in western Indonesia, whereas *Eucalyptus* trees occur mainly in the east.

Vegetation types are controlled by soil, climate and altitudinal characteristics, of which many different types exist in Indonesia. Lowland rain forest, where seasons are indistinct and rainfall high, is the jungle of legend and occurs below 1,500 metres. More plant species are represented here than in any other vegetation type. The largest remaining areas are in Kalimantan and Irian Jaya. Elsewhere, the forest cover is rapidly becoming fragmented, especially in Sumatra. Montane rain forest occurs above 1,500 metres, in regions that are wet all year round. Such forests are relatively poor in species. Extensive montane rain forests occur along the Barisan Range of Sumatra and, also, in Irian Jaya, where the highest peaks rise above the upper limit of the forest.

Seasonal forests grow in areas with a distinct dry season, even though the total rainfall may be quite high. Such forests once covered most of Java and Nusa Tenggara but are now greatly reduced in extent. This southern region of Indonesia has a climate that is very arid during the middle months of the year when desiccating winds blow from Australia. These seasonal forest types do not support high levels of plant diversity. Few undisturbed areas remain because they are so susceptible to fire during the dry season.

Inland swamp forest occurs where the soil is more–or–less permanently inundated. It grows on either a mineral soil or on a layer of peat, which can be over 20 metres deep. The world's largest tropical peat swamp forests are in eastern Sumatra, southern Kalimantan and in Irian Jaya. Mangrove swamp forest grows on gently shelving coasts where currents are gentle. Indonesia's mangroves are among the most species–rich in the world, with those of Irian Jaya being second in size only to the vast Sunderban region of India and Bangladesh.

Lowland Rain Forest Formations

Lowland evergreen rain forest has the greatest diversity of species and most complex structure of any category of tropical rain forest. It also has the largest concentration of economically useful plants and is Indonesia's primary source of timber. Thus, while being of great scientific interest, it is also the most threatened of Indonesia's rain forest types. Lowland rain forests occur in many parts of the country.

A Wealth of Species

Lowland, evergreen rain forest is at its most luxuriant in Kalimantan, where a single hectare may contain over 200 species of tree with a diameter of ten centimetres or more, and roughly every second or third tree belongs to a different species, which means that a single square kilometre of rain forest in Kalimantan is likely to contain about 1,000 tree species. Elsewhere, the diversity of species is somewhat less. This is especially the case in places where there is a short, but marked, dry season. Here, the forests have a slightly different structure, and some of the larger tree species will be deciduous. These forests are identified as semi-evergreen rain forest.

»»Limestone cliffs in the Maros Hills, Sulawesi.

Distribution and Regional Differences

The main blocks of Indonesia's rain forests are centred on Kalimantan in the west and Irian Jaya in the east. A north to south belt runs through Sulawesi and Maluku in between which there are areas of more seasonal climate that bear monsoon forests. There are a number of significant differences between the two regions. In Irian Jaya, for example, a general instability of the land due to frequent earthquakes and landslides on the steep, immature mountain slopes, has had a profound influence on the structure and composition of the forest, although these effects are less obvious the further east one travels.

Heath forest (Karangas) along the Sekunyer River in Tanjong Puting, Central Kalimantan.

Besides local differences in the floristic composition, there are also major differences in lowland rain forests in west and east Indonesia. One of the most important of these is the predominance, in the west, of the tree family Dipterocarpaceae. This family is not only abundant in numerical terms, but it is extremely rich in species, with ten genera and 386 species. No other tree family has comparable levels of diversity in the world's forests.

Dipterocarps occur in all kinds of lowland and lower montane rain forest, but they are at their most diverse, in terms of species numbers, in lowland, evergreen rain forest where, in parts of Borneo and the Philippines at least, they dominate the upper canopy. These dipterocarp forests have a grandeur and stature not found anywhere else in the humid tropics. This, together with their wealth of plant and animal species, places them at the apex of evolution on Earth. In eastern Indonesia, however, it is noted that there are far fewer dipterocarp species and they are not commonly found in the rain forest. Here the rain forest is generally shorter in stature and less species-rich than forests growing in the western region. There are also fewer emergent trees and when seen from the air the canopy gives the appearance of being much smoother.

»»Buttresses of Heritiera simplicifolius in the lowland rain forest of Gunung Palung National Park in Kalimantan.

Heath Forest

Sandy soils are extremely acidic and usually podzolised — that is, with a hard iron or humus 'pan' layer at some depth in the soil profile. These soils support a very special forest formation, widely known as heath forest or *kerangas*. This is typically a dense forest of pole-like trees, lacking in emergent species, and with a rather even canopy crown of a greyish-green hue. The average leaf size is smaller than in other lowland forests and the forest floor is brightly illuminated owing to the considerable

VEGETATION TYPES AND BIOGEOGRAPHY

DOMINANT RAIN FOREST TREES
The family Dipterocarpaceae takes its name from its characteristic two–winged seeds. Dipterocarps are typified by large leathery leaves and brightly coloured flowers. They can grow to heights of 70 metres and more, with circumferences in excess of six metres. They tend to flower infrequently, but synchronically. Flowering seasons are probably triggered by periodic droughts which occur every five to seven years, and are followed by mass fruiting seasons: it seems that the glut of oil–rich dipterocarp fruit satiates the appetites of fruit–eating animals, allowing a good number of seeds to survive their predations. Dipterocarp seedlings are found in great abundance on the forest floor where they are able to tolerate the very low light levels until a natural break in the canopy, caused by a falling tree or landslip, provides an opportunity for the most vigorous saplings to claim their place in the forest crown.

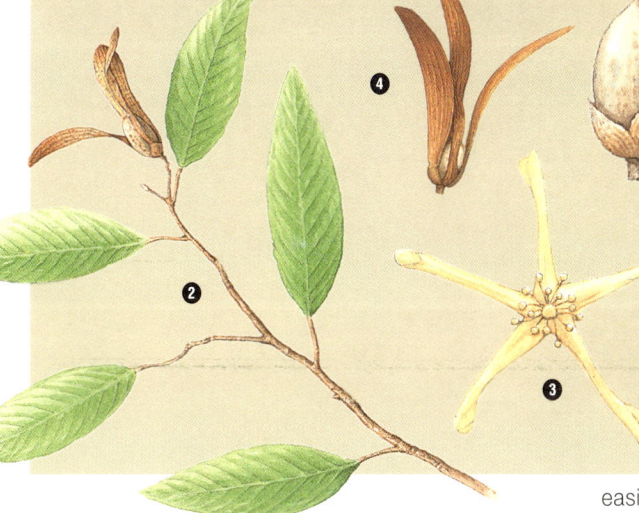

Shorea curtisii *Dyer ex King*, a dominant rain forest tree:
1. *Tree habit*
2. *Fruiting branch*
3. *Corolla*
4. *Fruit*
5. *Nut*

penetration of light at ground level. Most heath forests are liable to drought because of the coarse–textured, freely draining soil, but it is not yet known to what extent their highly distinctive features are an adaptation to withstand occasional water stress. Virgin heath forest soils are no poorer in mineral nutrients than other forest formations, but are very easily degraded once the forest is disturbed. Myrmecophytes (ant plants) and pitcher plants grow abundantly in heath forests and are specialised to supplement their intake of nutrients from the soil.

Heath forests are drained by tea–coloured streams bearing humus in suspension. The greatest concentration of these forests is in southern Kalimantan, but they also occur scattered elsewhere across the Archipelago.

Limestone and Serpentine Forests

Forests growing over limestone and serpentine rocks are distinguished by species that are able to tolerate the conditions of these sites. Some special limestone sites, for example, are especially prone to periodic desiccation and plants that grow in such locations show adaptations which enable them to withstand this. Limestone occurs either as tower–like outcrops, with almost vertical sides, or else as conical hills. These are found scattered through Kalimantan and Sulawesi, but are most extensive in Irian Jaya. In the latter instance, however, the influence of the rock on the forest is often masked by deep overlying layers of clay soils.

Serpentine rocks, which are rich in heavy metals and contain an unusually high magnesium content, are very extensive in east–central Sulawesi, but also occur elsewhere.

Freshwater Wetlands

Indonesia has vast and remarkably diverse wetlands, which cover 37 million hectares or nearly one-fifth of the country's land mass. They range from land that is periodically flooded to areas that are permanently waterlogged with fresh, brackish or marine waters. Freshwater wetlands include freshwater swamp forest, peat swamp forest, herbaceous swamps and lakes. The largest wetlands are found in the low–lying alluvial plains fringing Sumatra, Kalimantan and Irian Jaya.

»» *The swamp forests of Rawa Danau Reserve in West Java are flooded each year. The last of their kind in Java, they botanically resemble those of Kalimantan and Sumatra.*

Swamp Forests

Swamp forests are forests which are regularly flooded by fresh water. In permanently wet sites, organic matter accumulates and peat is formed, leading to the development of peat swamp forest. Flooded forests without peat are simply called freshwater swamp forest. Indonesia's freshwater and peat swamp forests, which once covered about 10 and 20 million hectares respectively, are the largest examples of these habitats in Southeast Asia. Indonesia's peat swamp forests are the largest expanse of this habitat type in the world, as there is little evidence of peat development in the lowlands of South America or central Africa. Much of this swamp forest has been converted to rice paddies or other forms of agriculture. On Java, almost all swamp forests have disappeared, while on Sumatra and Kalimantan undisturbed swamp forest is now rare. Swamp forests are a source of drinking water, wildlife habitats, timber, and products such as *jelutung* (the latex of *Dyera costulata*) which is used in paints and chewing gum.

Many swamp forest trees have special adaptations to waterlogged conditions. Stilt roots provide extra support in the soft ooze, while pneumatophores, or air roots, enable roots to breathe, oxygen being in short supply in flooded soils. Species diversity is low in comparison with lowland rain forest ecosystems, but the swamp habitat contains numerous unique, curious and often useful species.

There are many types of swamp forest — peat depth and inundation levels being important factors in determining the type of forest formation and local ecology. Shallow–to–moderately deep peat supports tall, relatively species–rich, mixed swamp forests. Berbak National Park in Sumatra is a good example, with 175 tree species recorded to date, dominated by *Alstonia pneumatophora*. which grows up to 30 metres in height. In other parts of Sumatra, and in Borneo, where the peat is very deep, an impoverished, uniform and open–canopied forest prevails, domixnated by *Calophyllum* spp. and *Pandanus furcatus*. This type of forest formation is known as padang peat swamp forest; here the water appears almost black. Such forests also occur in areas where there are great variations in the water level during the course of the year, while stunted forests, dominated by *Barringtonia acutangula* and *Syzygium* spp., are typically found in deeply flooded swamps such as those which occur in parts of Kalimantan.

A particular locality may have a wide variety of forest types each associated with different degrees of inundation. At Danau Sentarum, in West Kalimantan, for example, 228 tree and shrub species have been recorded in swamp forests, although individual vegetation types may only have 12 to 50 species.

Towing bundles of sedges, Lepironia articulata *home on Lake Air Hitam in South Sumatra, for use in the cottage industry of mat–making.*

*Herbaceous swamps such as this one dominated by lotus (*Nelumbo nucifera*), water lettuce (*Pistia stratiotes*) and* Ludwigia adscendans, *are often secondary vegetation types. These develop where clearing and burning have removed the original swamp forest.*

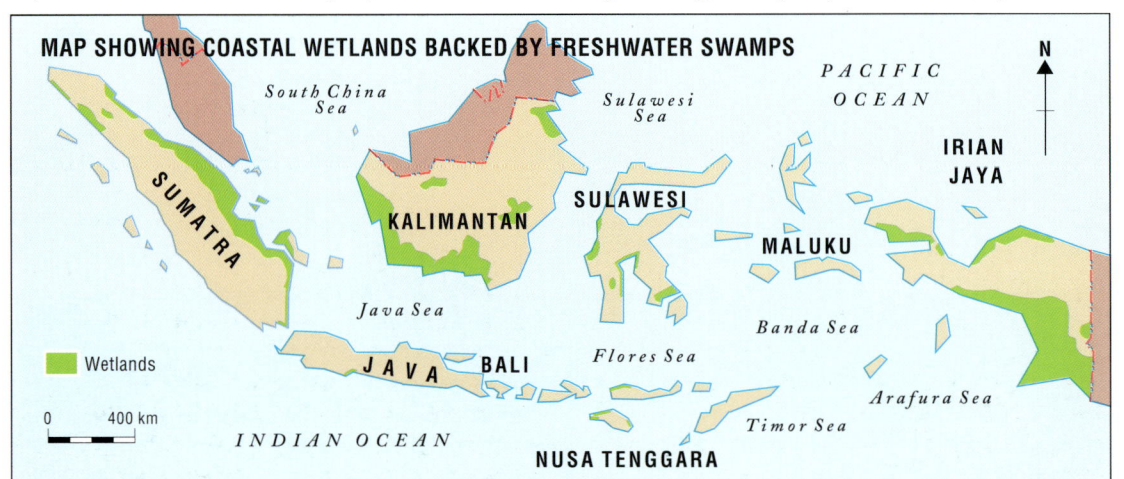

VEGETATION TYPES AND BIOGEOGRAPHY

cent of the total area of Indonesia, they are very important as water resources, fisheries and tourist attractions. Many plant species occurring in herbaceous swamps are also found along lake margins or, in the case of nutrient-rich lakes, in floating mats of vegetation. More typical of the lake environment are a variety of submerged, semi-submerged and free-floating aquatic species. These include fully submerged species such as *Hydrilla verticillata* and *Ottelia alismoides*; submerged species with floating leaves such as the water lily *Nymphaea nouchali*; suspended species such as bladderworts (*Utricularia aureum*); and free-floating species such as the Nile Cabbage (*Pistia stratiotes*) and water fern (*Salvinia molesta*). Most of these plants are common throughout Southeast Asia.

In contrast to other wetland habitats the lakes have relatively few endemic species, which reflects their relatively recent origin. Exceptions are *Ottelia mesenterium*, (Lake Matano-Towuti in South Sulawesi, and *Najas marina* ssp. *sumatrana* (Lake Maninjau in West Sumatra). Lake shores are favourite sites for human habitation, which probably accounts, at least in part, for the widespread dispersal of exotic species. A recent review of Indonesia's 35 major lakes and lake systems showed that exotic plants have been introduced to 75 per cent of the lakes studied, while exotic fish are found in 90 per cent of cases. Exotic weeds, especially water hyacinth (*Eichhornia crassipes*), are now a serious problem on many of Indonesia's lakes affecting fisheries and disrupting transportation.

A pneumatophore — a submerged or exposed root which also acts as the respiratory organ of the plant — in a peat swamp forest in West Kalimantan.

«« *The paper bark tree,* Melaleuca cajuputi *seen in a plantation in Sumba. The detail shows the light grey bark which tends to peel off in layers in old trees and gives protection against fires. Kayu putih oil is distilled from the leaves.*

In some instances swamp forests have been formed by human activity. For example, clearing and burning has created vast *Melaleuca cajuputi* (paper bark) swamps in Lampung (Sumatra), South Kalimantan and southeast Irian Jaya, while managed sago (*Metroxylon sagu*) swamps are widespread in South Kalimantan, Irian Jaya, South Sulawesi and Southern Aceh (Sumatra).

Herbaceous Swamps

In the past, herbaceous swamps only occurred at high altitudes, above the tree line, or else in dynamic habitats such as along streams or shallow, infilling lakes. Logging, clearing and burning, however, have considerably increased the expanse of freshwater herbaceous swamps, at the expense of swamp forests. Herbaceous swamps are especially common in South Kalimantan (Negara River swamps) and South Sumatra (Ogan-Komering swamps). This covers a total area of more than two million hectares, and researchers have recorded more than 600 herb species, with sedges, true grasses and members of the marigold family dominating. Thirty-nine of these swamp species are endemic to Indonesia, while 119 have been introduced from elsewhere. Many of the latter are a nuisance in rice fields and difficult to eradicate. Water hyacinth (*Eichhornia crassipes*), a noxious weed from Amazonia, was first cultivated in the Kebun Raya in Bogor in 1894, from where it escaped via the Ciliwung River and spread throughout Southeast Asia.

Inland Water Bodies

Indonesia has hundreds of lakes which are the legacy of tectonic, volcanic or glacial activities in the region, or are the result of flooding in the plains of wetter parts of the Archipelago. Although their combined surface area represents less than 0.25 per

TYPICAL PEAT SWAMP ON SHALLOW PEAT FOREST ON SHALLOW (LEFT) AND DEEP (RIGHT) PEAT

A An uneven-canopied high forest similar in structure and physiognomy to lowland dipterocarp forest but with fewer species per unit area, a lower canopy (36–42 metres) and fewer stems per unit area.

B A dense, even-canopied forest with xeromorphic and stunted features, with few trees exceeding 1.8 metres girth.

Source: After Anderson 1961.

A trend of decreasing fertility from the perimeter to the centre of the convex peat dome is demonstrated by the above sections of vegetation found on thin (<2m) **A** and deep **B** (>2m) peat. There is a decreasing trend in number of species, canopy height, total biomass per unit area, girth of individual trees from thin to thick peat. This is accounted for by the decrease in the amount of inorganic nutrients in soil towards the centre of the swamp, particularly phosphorous and potassium.

Coastal Wetlands

Indonesia is extremely rich in wetland habitats, of which coastal wetlands form an important part. These include mangroves, mudflats and seagrass beds. Encompassing some 2.5 million hectares, Indonesia's mangroves are the most extensive and biologically diverse in the world.

Tangled Forests

Mangrove forests occur in flooded marine and brackish environments in the tropics and subtropics. Mangrove species are specially adapted to flooded habitats, high salinity and muddy, unstable soils. Being tolerant of saline soils, they are halophytes. As with freshwater swamp species, they often have stilt or prop roots, and pneumatophores. Pneumatophores are specialised 'breathing roots' containing pores which are found in plants that are adapted to living in waterlogged soils where oxygen is in short supply. Other adaptations of mangrove species include the ability to remove salt from their tissues, for example, by their leaves excreting salt.

Many typical mangrove tree genera, such as *Avicennia*, *Bruguiera*, *Ceriops* and *Rhizophora*, have seeds which germinate while still attached to the tree, and what is commonly perceived to be a long *Rhizophora* fruit is in fact an already germinating plant. This adaptation enables the seedling to become rapidly established once it falls from the parent tree.

Indonesia has more mangrove forests than any other country in the world. This habitat once extended over more than four million hectares but, by 1990, this area had been reduced by 40 per cent, mainly through land clearance and conversion to agriculture or ponds for aquaculture. More than half of the remaining mangroves occur in Irian Jaya, with only a small area remaining in Java.

Contrary to popular opinion, mangrove forests support high levels of biological diversity. At least 200 species of mangrove seed plants and ferns, including 80 trees and shrubs, have been recorded to date from Indonesia's mangroves. Worldwide, about 60 species occur exclusively in mangroves: the 'true mangrove' species. The highest concentration — 40 species — is found in Indonesia, while mangrove forests of Africa and America have less than 10 true mangrove species. Because of the ease of water-borne dispersal, few species are unique to Indonesia. Exceptions are *Ixora timorensis*, which occurs only in Nusa Tenggara, and the rare mistletoe, *Amyema anisomeres*, which is only known from one mangrove area in South Sulawesi.

Mangrove vegetation typically displays zonation patterns which have been linked to soil type, exposure, salinity and tidal influence. Indonesia's

View of a Rhizophora dominated mangrove on Komodo island in Nusa Tenggara.

DISTRIBUTION OF REMAINING MANGROVE FOREST

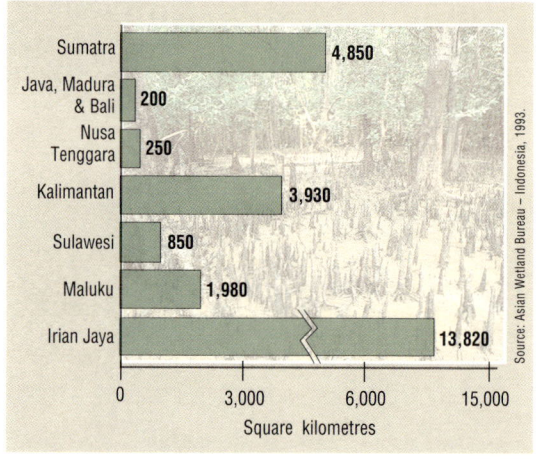

Region	Square kilometres
Sumatra	4,850
Java, Madura & Bali	200
Nusa Tenggara	250
Kalimantan	3,930
Sulawesi	850
Maluku	1,980
Irian Jaya	13,820

Source: Asian Wetland Bureau – Indonesia, 1993.

THREE DOMINANT ROOT SYSTEMS IN MANGROVE SPECIES

The respective root systems illustrated have evolved in adaptation to the various conditions in the mangrove habitat.

❶ Spike roots typical of both *Sonneratia* and *Avicennia* spp.
1a. An erect breathing root of *Sonneratia* spp.

mangrove trees may range from dwarf *Avicennia marina* stands, which grow to the height of one to two metres on the seaward side of accreting shores, to 30 metre high mixed forests of *Bruguiera*, *Rhizophora* and *Ceriops* spp. On more exposed, but otherwise stable coastlines species such as *Sonneratia alba* and *Avicennia alba* grow, while in waters of lower salinity, such as estuaries, *Nypa fruticans* and *Sonneratia caseolaris* are common. Apart from saplings, the undergrowth is usually sparse, although *Acanthus ilicifolius* and the mangrove fern, *Acrostichum aureum*, may grow along the banks of creeks and in disturbed areas.

Mangroves play an important part in the ecology of marine and coastal environments, protecting shorelines from erosion, and providing food, shelter and breeding areas for fish and crustacean species, many of which have commercial value. In some regions there appears to be a direct correlation between the productivity of offshore fisheries and the presence of mangrove swamps in the locality: where these have disappeared, fisheries have suffered.

Beds of Seagrass

Seagrasses are light–demanding plants and grow in shallow, clear waters, although some species such as *Halophilia ovalis* may occasionally occur at depths of 30 metres. They differ from seaweeds in that they are seed plants which are adapted to living in marine and brackish environments. Only a limited number of seed–plant families have colonised these habitats and, of the 50 species which are known to exist worldwide, 14 are found in Indonesia.

Even in nutrient-poor waters, seagrass beds are very productive environments, because seagrass, unlike algae, can extract nutrients from marine sediments. At the same time, seagrass is able to fix dissolved atmospheric nitrogen in a process similar to that which occurs in legume root nodules. As they are higher plants with extensive root systems, seagrasses are restricted to sand and mud habitats. Here they form an important habitat for crustaceans, marine turtles, fish, molluscs and dugong. Some of these feed directly on the living plants, while others graze on their epiphytic algae. Decomposing seagrass forms the basis of an extensive food web.

The role of seagrass beds is often overlooked, but these formations fulfil an important ecological role in that they help stabilise soft, drifting sediments. This can be an important consideration in situations where shifting sediments might adversely affect coral reefs, beaches or shipping channels.

Enhalus seagrass beds (foreground) growing in their natural habitat. Mangrove trees Sonneratia alba with spike roots in the background, reflect the diversity of forms in both groups of angiosperms.

Halophila ovalis, *a seagrass often found along the Indonesian coast. Its features are:*
A *Paired leaves with oblong–elliptic blades*
B *Flower of female plant*
C *Roots firmly adhering plant to the substrate*

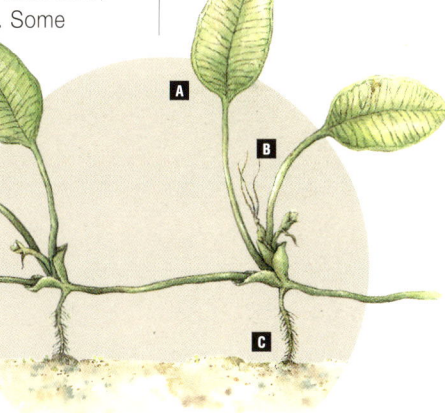

❷ **Knee roots** of *Bruguiera* spp.
2a. *Bruguiera* branch bearing seedlings.

❸ **Stile roots** of *Rhizophora* spp.
3a. *Rhizophora* propagule hanging from tree.

RHIZOPHORA SEEDS

Rhizophora seeds germinate while still on the tree. This enables a plant to quickly establish itself in a dynamic environment. The elongation of the fruit (3a.) is actually the developing hypocotyl (3b.), or primary stem, on which the roots (3c.) develop. The torpedo–like shape enables the propagule to penetrate the soft mud upon falling from the tree (3d.).

Montane Forests

The moist, cool highlands of Indonesia support forests which, though less complex than evergreen lowland forests, nevertheless contain a very rich, fascinating and unique array of plants. Large areas of montane forest have been cleared for agriculture, or are maintained in a degraded condition by fire, but those that remain constitute an important refuge for many plant and animal species.

»»*Pandanus moss forest in Gunung Tujuh National Park, Sumatra.*

Altitude Zones

As one travels from the coastal lowlands to highland areas in Indonesia, the air temperature drops at an average rate of about six degrees Celsius per 1,000 metres. These cooler climates are less favourable for plant growth and the natural forest, though still dense and varied, is progressively reduced in stature. There is also less diversity in plant species the higher one climbs. These transformations in the structure and composition of the forest are continuous and vary between regions, so it is difficult to define precisely the altitude at which one type of forest is replaced by another. In Indonesia's largest mountain region, Irian Jaya, forest structure and composition both tend to change noticeably at about 1,200 metres, which is usually taken to mark the lower limit of montane forest. Another change, towards shorter, shrubbier forest, with even fewer species, occurs around 3,000 metres, marking the transition from lower to upper montane forest. The plants which grow at these altitudes tend to be quite specialised in terms of their adaptation to a colder climate. Elsewhere, vegetation limits vary according to local circumstances and topography. For example, upper montane forest occurs at lower altitudes on smaller mountains.

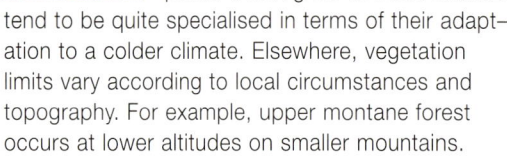

Although a greater incidence of cloud cover at high altitudes reduces the overall level of sunlight, the atmosphere is naturally thinner than at sea level and therefore absorbs less of the damaging ultraviolet-B radiation present in direct sunlight. This increased intensity of sunlight may explain some peculiarities of montane forest trees, such as thick leathery leaves, and the tendency for young foliage to be pink because of protective substances within it. Vegetation types are also determined by local soil conditions. Rich soils support taller, more diverse forests than soils which are poor in essential plant nutrients, such as those developing on recently erupted volcanic deposits.

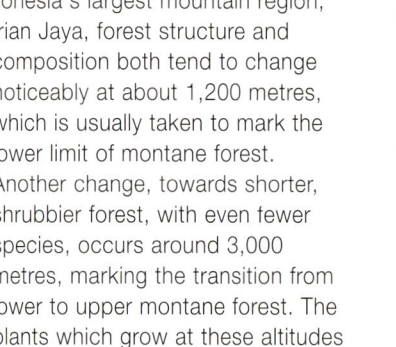

A lichen–draped tree growing in moist conditions, at an altitude of about 2,000 metres, in the montane forest in the Baliem Valley, Irian Jaya.

Montane Forest Flora

The montane forests of Indonesia are by international standards very diverse. Their floral composition not only varies with altitude, but also differs from one region to another. Beech trees (*Nothofagus* sp.), for example, are a characteristic feature of montane forests in Irian Jaya but are not found further west. They do, however, have affinities with related plant species in the cool, southern rain forests of Australia, New Zealand and South America. This is because they all share a common origin with the ancient rainforests of the great southern supercontinent Gondwana, which, over the last 100 million years, fragmented into the present separate southern continents. The island of New Guinea was once part of Gondwana, but more western parts of Indonesia have a different geological origin, namely

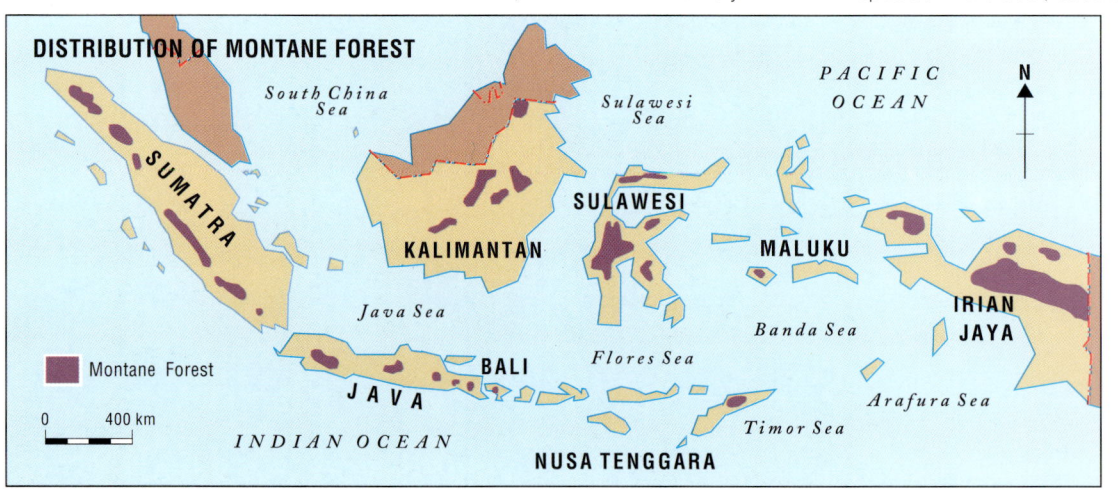

VEGETATION TYPES AND BIOGEOGRAPHY

Epiphytes and Mosses

Epiphytes flourish in the humid montane forests of Indonesia. Indeed, the highest montane forests have such an abundance of epiphytic mosses that they are sometimes called moss forests. Other epiphytes are also common, ranging from small, simple plants like algae and lichens, to larger species such as ferns and orchids. In some instances a tree may become so well endowed with epiphytic species that it will actually collapse under their weight.

Moving through the upper montane forest where there are no paths is a slow and exhausting business. There are no tall tree trunks to walk between and instead one must clamber through a tangle of branches. Everything, from the trees to the ground itself, is covered by a soft green blanket of moss, making it difficult to know whether it is solid rock, slippery wood, or a hollow underfoot. Grasping a branch for support can be a risk — it may be alive and strong, but equally it could be rotten and weak.

Despite the fact that there are many trees and epiphytes in a montane forest, one still finds quite a few plants growing on the shady forest floor, ferns and wild gingers among them. In the highest and most stunted montane forests, more light penetrates through the canopy and some alpine plants may be found, along paths, beside streams and in places where the tree cover is broken. Climbing, carnivorous pitcher plants are also a feature of such open places.

On the forest floor of Gunung Gede near Cibodas in West Java, the bright pink flowers of the Impatiens *sp. vie for attention.*

The Value of Montane Forest

Montane forests are an invaluable resource for Indonesia. Unfortunately large areas have been cleared for farming or forestry and, although in some places these activities have been abandoned, the forest fires often prevent the regeneration of vegetation. In East Java, such burned areas often become dominated by a single species of tree, *Casuarina junghuhniana,* which grows over a readily combustible grassy understorey. Elsewhere, surviving montane forests continue to provide traditional foods and valuable timber. With care these uses can be sustained, to provide further resources for future generations. These forests also provide opportunities for healthy outdoor recreation for Indonesia's increasingly urbanised population. Montane forest has a very high conservation value and international tourism potential, as in Irian Jaya where it supports special fauna such as the exotic and colourful birds of paradise.

Young fern fronds on the forest floor of cloud forest, found 2,500 metres above sea level in the Baliem Valley, Irian Jaya.

the northern supercontinent of Laurasia. Similarly, there are plant species found in western Indonesia which are scarce or even absent from Irian Jaya. The Indonesian maple, *Acer laurinum*, whose range extends as far northwards as Thailand, but which is not found east of Sulawesi, is one such example. Other plant species, however, are more widely distributed. These include such varied and common groups as the oaks (*Lithocarpus*), figs (*Ficus*) and tree ferns (*Cyathea*).

The tall trees of lower montane forests provide a very diverse, dense canopy. Most of the botanical interest lies well above the ground, and may be hard to see, especially when mist hangs in the still, cool air between the trees.

In the upper montane forest, conifers like *Podocarpus,* are commonly the tallest trees present, their dark crowns emerging above the surrounding canopy of broadleaf trees, providing a beautiful sight when seen against a background of white cloud. At these higher altitudes, trees are neither as tall or as densely packed together as in lowland forest formations and more light is able to penetrate to the forest floor where a variety of small flowering plants can be found. The large blue gentians (*Gentiana lateriflora*) of Sulawesi are particularly beautiful, while the pink or orange flowers of balsam (*Impatiens*), which occurs in most montane forest, may brighten wet places near rocky streams and waterfalls. White and mauve violets (*Viola*) are also widespread, and are often conspicuous beside paths.

RHODODENDRONS

Rhododendrons are a particularly colourful feature of montane forests. These shrubs occur in great variety throughout Indonesia's highlands, with different species being found at different altitudes. In the cool wet conditions of the upper montane forest, rhododendron species have red flowers to attract nectar–seeking birds. While those which grow at lower, warmer altitudes, usually have either white or pink flowers with a sweet fragrance, and are pollinated by nocturnal moths, or else bear orange blossoms and are pollinated by day–flying insects such as butterflies. Indonesian rhododendrons are grown for their beautiful blooms by keen gardeners all over the world.

Alpine Vegetation

At high altitude, tree growth is retarded and forests give way to alpine vegetation including tundra, tussock grassland, and a variety of bog, crevice and heathland communities. Most of the small plants growing in the alpine vegetation belt are related to species found in other cold climates around the world, and have no close affinities with forest plants at lower altitudes.

»»*Alpine meadows near Lake Discovery in Irian Jaya, located 4,000 metres above sea level — alpine grassland with prominent stands of tree ferns are seen in the foreground .*

Alpine Vegetation Types

The alpine belt in Indonesia lies above an altitude of about 3,800 metres where it is too cold to permit the growth of trees. The largest area is in Irian Jaya, where the highest mountain, Gunung Jaya (4,884 metres), a soaring limestone peak, is located. Snow covered, it overlooks small glaciers, and extends into frigid desert, above the limit of all vegetation.

Alpine vegetation varies between regions, with altitude, soil type and drainage. In Irian Jaya, a low-growing tundra vegetation, with mosses and lichens, but few flowering plants, survives at high levels, up to about 4,700 metres. Tussock grasslands occur more extensively at lower levels. These dominating the latter vegetation may be a metre tall, with long, narrow leaves and flowering stalks growing from pedestals of dead stems bound together by roots. Old, dead leaves remain attached to the plants for years, outnumbering living, green leaves and lending the tussock grasslands a pale brown colour.

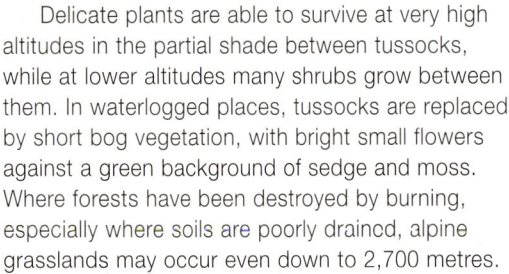

Delicate plants are able to survive at very high altitudes in the partial shade between tussocks, while at lower altitudes many shrubs grow between them. In waterlogged places, tussocks are replaced by short bog vegetation, with bright small flowers against a green background of sedge and moss. Where forests have been destroyed by burning, especially where soils are poorly drained, alpine grasslands may occur even down to 2,700 metres.

Although the climatic forest limit lies at about 3,800 metres, alpine plants may also occur on smaller mountains throughout Indonesia, in some cases at 2,000 metres or lower, wherever tree growth is stunted or absent. Shrubby alpine heathland grows on exposed, rocky ridges, and short herb vegetation is found in bogs and beside streams.

Javanese edelweiss, Anaphalis javanica, a common shrub in the high altitudes of Gunung Gede–Pangrango National Park. The small delicate white flowers make a sharp contrast with silver-coloured narrow, thick and hardy leaves.

Alpine plants may also find suitable habitats in disturbed places, such as mountain paths. The volcanic mountains of Java, Sumatra and Bali often have rather barren summit areas, with rocky or sandy ground where a few alpine plants survive in moist or sheltered places. Other mountains are completely clothed by forest, but still have some alpine plants growing beneath the shrubby summit forest where sufficient sunlight penetrates.

Microclimate

Alpine vegetation varies between different parts of the same mountain as a result of local climatic variations. The colder climate at higher altitudes results in more stunted vegetation and fewer species.

Topography is influential in other ways. On calm nights the ground surface air cools, and flows downhill, collecting in flat valley bottoms. This leads (with waterlogged soil as an additional factor) to 'frost hollows' where plants typical of higher altitudes may grow. Alpine grassland in these hollows may be surrounded by forest on the warmer slopes above, forming an 'inverted timberline'.

The direction that slopes face also influences climate and vegetation. Strong prevailing winds stunt the vegetation on exposed slopes, as on Gunung Rantemario in Sulawesi. Generally, slopes facing east have warmer climates than those facing west, due to being warmed by the morning sun when skies are more often clear. In Irian Jaya this results in more species growing on east-facing slopes, than the opposite ones at the same altitude. In Java, some alpine shrub flowers open earlier on the eastern side.

Alpine Flora

There are many small but interesting plants in Indonesia's alpine flora. Yellow-flowered buttercups (*Ranunculus*) grow besides blue gentians (*Gentiana*) and purple or white violets (*Viola*) brighten the low turf of short grasses and sedges. The alpine landscape is also enlivened by beautiful shrubs. Splendid red-flowered members of the heather family (*Rhododendron* and *Vaccinium*) contrast with white-flowered species belonging to the magnolia (*Drimys*), tea (*Eurya*) and Australian heath (*Styphelia*) families. These are closely related to plants growing in other

CHANGES IN MOUNTAIN VEGETATION BY ALTITUDE

- Bare rock
- Tundra
- Grassland
- Open shrubby montane forest
- Closed montane forest

Metres: 5,000 / 4,000 / 3,000 / 2,000

Source: Montane forest of Indonesia

VEGETATION TYPES AND BIOGEOGRAPHY

FURRY LEAVES
Many plants in Indonesia's sub-alpine and alpine vegetation have leaves covered in dense hairs such as *Potentilla* sp. (right). They belong to several different families, although the daisies are particularly well represented. The hairs may have evolved as a protection against damaging levels of solar radiation which at high altitude are not filtered by the full depth of the earth's atmosphere. Such intense radiation can also heat exposed leaves, leading to rapid and potentially damaging water loss through evaporation. A furry but reflective covering on the leaves not only reduces the extent to which leaves become heated but also reduces the air flow around them, thereby decreasing the rate of evaporation.

Some plants commonly found in alpine vegetation, clockwise, below: Representatives from the herbaceous habit include: Gentiana sp. with their tiny blue flowers. Yellow-flowered buttercups, Ranunculus sp. are usually found growing next to gentians. The alpine rhododendron, a shrub with bright fiery red flowers such as Rhododendron saxifragoides, adds a splash of colour to the landscape.

cold places around the world, but in Indonesia they occur in unusual combinations. Some are typically northern hemisphere plants, such as the gentians and rhododendrons, while others are typically southern provenance as in the case of *Styphelia* and the daisy bushes (*Olearia*). It is only in these mountains that such species are found growing side by side.

Some of these alpine plants are unique to the region. An alpine shrub (*Detzneria*) with deep blue flowers that never open, is endemic to Irian Jaya. Other alpine plants are so rare, or grow in such inaccessible places, that botanists know them from only one or two specimens which were collected many years ago and have never been seen since.

The daisy family is one of the most important and diverse groups in Indonesia's alpine vegetation. At the highest altitudes, there are tiny, sessile daisies with hardly any stem, while lower down the same mountain, taller varieties provide a splash of colour in sheltered gullies and along stream banks. Others grow as tall shrubs at the forest edge. Some daisies have large, pretty flowers; others lack petals and are a brownish-green. Diverse alpine grasses — some of which are restricted to a single mountain — range from large tussock forms to small creeping varieties. The heather family is well represented in Indonesia's alpine flora. Most are tough shrubs with red, pink or white bell-shaped flowers, but some varieties climb over other plants, creep along the ground or form low cushions in boggy places.

International Comparisons
Indonesia's alpine flora is adapted to an environment where plants grow all year round in a perpetually cool climate not subjected to the rigours of winter. Such a lack of seasonality distinguishes these alpine zones from similar habitats outside the tropics. Despite this, many alpine plants would not be entirely unfamiliar to botanists from other mountainous countries, being more closely related to plants found in remote cold places, than to plants growing in nearby forests. Part of the explanation may be that migrating birds carried seeds between distant mountains during the ice ages, the most recent of which ended about 12,000 years ago.

The Importance of Alpine Vegetation
Indonesia's alpine vegetation has not been seriously affected by human activities because it grows in rugged and remote locations, where the cold climate precludes agriculture. Mining operations have had a destructive effect in some parts of Irian Jaya, while burning has imposed changes on larger areas over a longer period of time. However, most of this vegetation remains in a natural state, wherein lies its main potential — tourism. People are attracted by the challenge of mountain climbing, and are appreciative of the alpine environment. Indonesia can develop a flourishing tourism industry based upon the high mountains, with alpine vegetation providing an incentive for people to visit these regions. Sufficient attention, however, should be given to protect these unique habitats from potentially adverse effects of unplanned tourism or similar development.

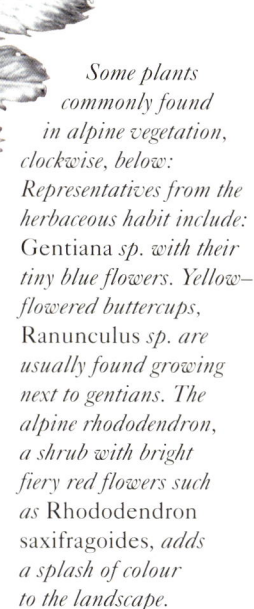

Monsoon Forest

Despite the common view that the tropics are continually wet, some parts of Indonesia have distinct dry seasons. The forests which occur in these drier regions are less diverse in terms of species than the ever-wet rain forests, but they are nonetheless the source of many economically valuable plants species. These monsoon forests are one of the most rapidly disappearing habitats in Indonesia today.

>> *Sandalwood (Santalum album L.), indigenous to Indonesia, valued for its fragrant wood and aromatic oil, has been on the endangered list for a long time. It is rarely found in the wild now.*

Epiphytic ant plants, Dischida sp. found in the seasonal forest of Flores.

SANDALWOOD (*Santalum album*):
A Fruiting branch with opposite leaves, purplish brown flowers and blackish red fruits
B Small tree which is a parasite of different plants in its juvenile stage

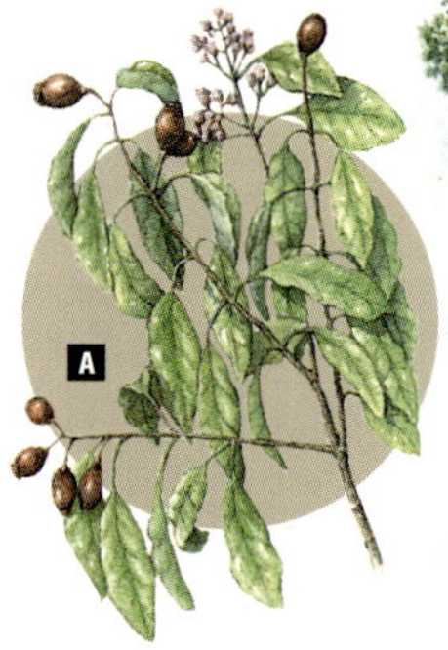

What Makes a Monsoon Forest?

Monsoon forests contrast with rain forests in that the sheer exuberance of vegetation found in a rain forest is missing in these seasonally dry forests. The diversity of tree species declines with increasing aridity while a lower, more open, canopy results in a greater number of deciduous species as opposed to evergreens. Fewer palms are represented and many species of damp-loving climbers, ferns, and orchids disappear altogether. At the same time, the under-storey of evergreen shrubs gives way to mixed herbs and grasses, and there is an increase in the occurrence of bamboo. Pure stands of single tree species are frequently found in monsoon forests. This may be due to natural gregariousness, but is often attributable to local peculiarities of soil or human interference.

Botanists tend to distinguish several monsoon forest types, ranging from tropical, moist deciduous forest, through dry, deciduous, to thorn forest. These are not distinct communities, but constitute a mosaic, where the degree of aridity, and local differences of soil and topography are vital influences.

Where they Occur

The distribution of monsoon forest in Indonesia forms a patchwork of forest areas interspersed by other vegetation types. Little original monsoon forest now remains outside protected areas. In eastern Java the best example is in Baluran National Park, but even here it is set in a mosaic of fire-climax grasslands. Other forests occur in Bali Barat National Park and to the east of Palu Bay, in central Sulawesi, where one disturbed remnant is left in the Paboya Reserve. Nusa Tenggara was once extensively covered by monsoon forest; today, the largest surviving tracts are in the Komodo National Park, lying between the islands of Sumbawa and Flores. In Maluku and southern Irian Jaya, there are still some areas, such as Buru Island, which have extensive whitewood or *kayu putih* (*Melaleuca cajuputi*) forests.

Dispersal of Species

How did monsoon forest species spread across the rainforest belts to their present positions in Indonesia? According to the famous tropical botanist, van Steenis, a lowering of the sea level during the Pleistocene Ice Age, and a corresponding increase in land surface, would have caused an expansion of the seasonal monsoon climate through western and central Indonesia. A fall in sea level of up to 200 metres meant that the islands on the Sunda Shelf —

DISTRIBUTION OF MONSOON AND SAVANNA FOREST

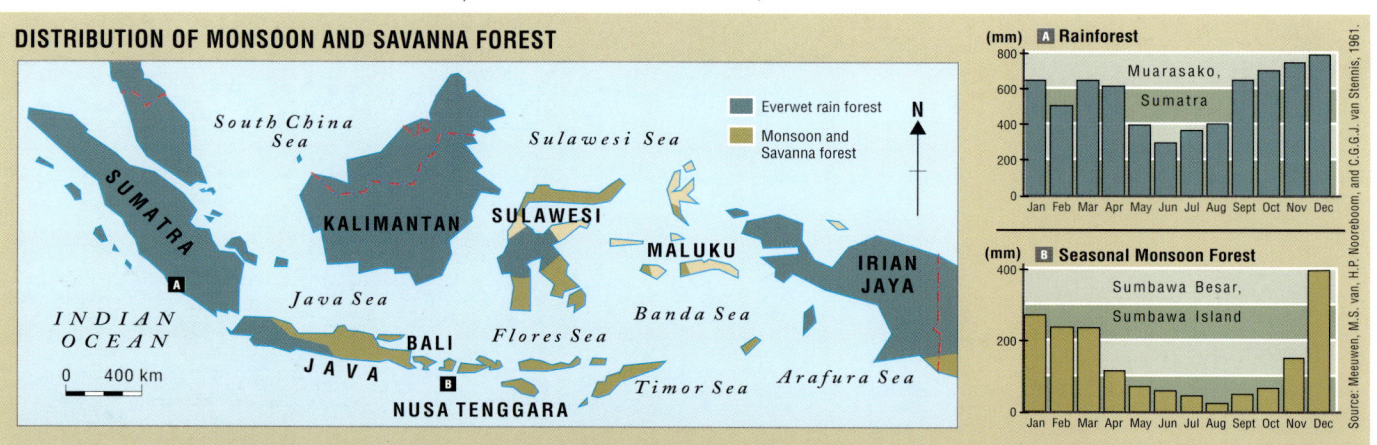

The monsoon forests of Indonesia have never been mapped precisely, but they are concentrated in a dry central band in Indonesia between the two big cores of ever-wet rain forest on the Sunda Shelf and in Irian Jaya. The graphs show the sharp contrast in rainfall between A a rain forest in Muarasako, Sumatra and B a monsoon forest in the region of Sumbawabesar, Sumbawa Island. A higher and more even rainfall distribution is experienced in the rain forest. Whereas in the seasonal monsoon forest, a lower rainfall and a marked difference in rainfall distribution is observed.

Sumatra, Java, Borneo and the southwest Philippines — became one huge land mass connected to mainland Asia by the Malaysian Peninsula. New Guinea and the Aru Islands on the Sahul Shelf were similarly united with Australia. When the Ice Age ended and the sea level rose again, these extensive areas of dry land shrank back to smaller refuges, while inter-island channels expanded, cutting off dispersal pathways. Van Steenis has suggested that the present range of some Leguminosea species are the legacy of this presumed extension of monsoon climate.

Useful Plants from Monsoon Forests

Little interest has yet been shown in the resources of monsoon forests. This is a surprising oversight considering that so many trees from Indonesia's monsoon forest provide timber, oils and food. Some of these important species are not actually native to the region, but have been introduced from the massive monsoon belt of mainland Asia. They include the neem tree (*Azadirachta indica*) from Burma, East Indian rosewood (*Dalbergia latifolia*), tamarind (*Tamarindus indica*) and teak (*Tectona grandis*). Tamarind and teak grow particularly well and have become naturalised in Indonesia. Teak was probably first introduced to the region by Indian navigators and merchants several centuries ago. Today these forests of central and eastern Java cover more than 700,000 hectares and teak is one of the leading commercial tropical timbers in Indonesia. Conversely, sandalwood (*Santalum album*), whose fragrant wood and aromatic oil has been highly sought after for centuries, is difficult to grow in plantations, while natural stands have almost been wiped out. Other valuable monsoon species include the Ceylon oak (*Schleichera oleosa*), source of the popular Macasar hair oil of the last century; *kayu putih*, which produces a highly-prized medicinal oil; and the candlenut tree (*Aleurites moluccana*) whose large, oily seed is used locally to make candles and commercially as a source of oil.

Fire and Shifting Agriculture

Dry herbaceous ground cover encourages fires, and recurrent burning furthers the spread of fire-resistant species such as tamarind, teak, *kayu putih* and *Casuarina junghuhniana*. The latter is especially resistant to fire and covers several thousand hectares in eastern Java. The fire-climax forests of *Pinus merkusii* in Sumatra are a similar example, while in Buru, where the grassy lowlands and hills are subject to a dry monsoon and are burnt annually, *kayu putih* is the dominant species. Although few grasses occur in many monsoon forests, repeated burning encourages their spread and kills off seedlings and small trees, thereby bringing about the permanent replacement of the forest with savanna grasslands.

Botanists now believe that many areas which should support monsoon forest are in fact covered by savanna. What is unclear, however, is whether if monsoon forest replaces rain forest and the trees become fewer and grasses more predominant, will savanna naturally replace this forest? Historical sources suggest that in some areas where, a century ago, monsoon forests grew, savannas now exists. But their history has not yet been studied rigorously and it is not clear how their replacement with savanna forests has come about. Are the areas so dry that only savanna and grasslands can exist naturally, or has repeated burning caused the original forest to open up? Were the fires started by lightening, or has the distribution of the monsoon forest been changed by humans?

Remaining original Acacia *monsoon forests of Gunung Baluran in East Java.*

*Pride of India (*Melia azedarach*), a dominant tree from the monsoon belt on the Asian mainland.*

Monsoon forest on the north coast of Wetar in the Lesser Sundas.

A Sumbanese craftsman carving sandalwood statues.

Savanna Forest and Grassland

Savannas are a common vegetation type in the driest areas of Indonesia. They are hunting and grazing lands, characterised by isolated trees standing in a park-like landscape, which is midway between grassland and forest. The question of their origins — whether these savannas occur naturally or are the result of human activities over the centuries — remains a subject of ongoing debate.

There are basically eight savanna types in Indonesia, which are found in two broad habitat divisions:
A. Soils more prone to water-logging:
❶ Zizyphus mauritiana and ❷ Tamarind (Tamarindus indica) savanna, a drought and fire-resistant vegetation which occurs from sea level up to 500 metres above sea level (asl.)
B. Drier soils:
❸ Palm savanna dominated by a. Borassus flabellifer or b. Corypha utan, found on rocky soil up to 400 metres asl.
❹ Melaleuca cajuputi savanna, usually found separate from other dominants near secondary swamp or on dry hills of low lying land
❺ Acacia leucophloea savanna on thick clay and limestone.

Characteristics

The presence of drought-resistant plants, dominated by mixed grasses and herbs, with scattered trees, means that the structure of savannas is quite different from that of rain forests. In the latter, where conditions are optimal for growth all the year round, most plants are either trees or epiphytes. In the case of savannas, however, seasonal droughts make continual growth impossible, and buds have to be protected from desiccation. Accordingly, one finds that most plant species are either annuals, which complete their life cycle from seed to seed within one wet season, or perennials whose above-ground parts die back during the dry season.

Although much anecdotal information on Indonesia's savannas exists, detailed, quantitative descriptions are completely lacking. Eight savanna types can, however, be identified, which are classified according to the dominant tree species (some of these may also occur in drier monsoon forests). Scant data are available on the composition of the ground cover, but in one study on Komodo, for example, the grasses *Themeda* and *Chrysopogon* dominate the undergrowth on dry, well-drained soils.

Distribution

Savannas occur in the same general area as monsoon forests. The vegetation in this region is essentially a patchwork of remnant monsoon forest, tracts of gallery forest or rain forest on south-facing slopes, and extensive savannas.

In the wetter areas of the Sunda and Sahul Shelves, localised rain-shadow effects create pockets of savannas. This occurs, for example, in Sumatra, around Padang Lawas and Pasemah; in Southeast Kalimantan in the Meratus Mountains; and in Northwest Kalimantan near Ranau. The Merauke area of Irian Jaya has similar vegetation while in the drier, central belt of Indonesia, savanna areas include the Palu Valley in Westcentral Sulawesi and all the islands of Nusa Tenggara. Other islands which have dry spots where savannas prevail include Buru, Seram, and Tanimbar, in Maluku, and also on Trangan, the southernmost island of the Aru group, which has sandy soils and experiences a seasonal climate.

Human Activities

Many of the most useful plants from savannas are also found in monsoon forests, but two species are worth special mention here. These are the *lontar* and *gebang* palms which are particularly prevalent on the very dry islands of eastern Nusa Tenggara. Savannas dominated by these palms are especially widespread on the islands of Timor, Rote and Savu, where they provide a crucial source of sustenance to the indigenous peoples during extreme droughts when no other source of carbohydrates is available.

The miracle of the *lontar* palm is that it produces copious quantities of sugary sap in the dry season. *Lontar* juice may be imbibed or consumed in a variety of forms: fresh from the tree, fermented, distilled, as a syrup or boiled down to a solid sugar. The Rotenese are even exporting *solid sugar* to Sweden. The leaves are used for roofing and weaving material, the leaf stocks for fencing, fibres for string, and the trunk for construction.

The *gebang* palm is an important source of starch which is stored in its trunk and processed like sago. Like the *lontar*, its leaf stocks are used for building materials.

Origins of Savanna

Botanists have long debated the evolution of savannas. One school of thought believes that they occur naturally as a result of climatic and soil factors which create an acute seasonal water shortage that inhibits plant growth. An alternative view is that savannas arise when forests are repeatedly burned. This may be the result of natural phenomena, but may equally be due to human intervention. No experimental data are available for the effect of different disturbances to the forest or the relationship between fire intensity and savanna communities. Nor do we know the potential for savanna grasslands to return to forest if the disturbances are removed. Historical records indicate that deliberate firing of grasslands is a long-standing practice in the region. In 1699, William Dampier observed that in Timor '... [the people] take but little pains to clear the land; for in the dry time they set fire to the withered grass and shrubs, and that burns them out a plantation for the next wet season. Their plantations are very (sparse) for they delight in hunting.'

The origins of these savannas are not just of academic importance. In many of the drier parts, people eke out a poor living from agriculture. One response to this situation would be to reforest these areas. If however, savannas occur naturally, then any attempts to reforest them may be a wasted effort.

The coarse grass, alang–alang (Imperata cylindrica) *is a very different but important vegetation type that confuses the debate on the origins of savanna.*

««*The savannas of Sumba provide grazing for the white ongole cattle.*

❻ Eucalyptus alba *found on the well drained lime-poor soil up to 700 metres asl.*
❼ Albizia chinensis *savanna and*
❽ Casaurina junghuhniana *both occur together in hilly regions up to 3,100 metres asl.*

Secondary Forest

If primary forest is cleared and the land then allowed to recover naturally, a completely different kind of forest, known as secondary forest, develops. Secondary forest has a simpler structure than primary forest and a distinctive set of tree species though these are considerably fewer in number than in the case of primary forest.

»» *When original primary forest is cleared and logged, a less species–rich secondary forest replaces it. This particular example is in Way Kambas, Sumatra.*

Development of Secondary Forests

In the rural landscape of Indonesia, secondary forest, or *belukar*, was once found mainly as the 'bush fallow' phase of the shifting cultivation cycle. Typically, a period of two to three years cultivation of hill rice, and other crops, would be followed by an interval of seven to 30 years fallow, when the forest was again allowed to regenerate and the soil regain its natural fertility.

Today, however, *belukar* occurs far more widely and in a variety of new contexts. Since the late 1960s, much of the Indonesian rain forest has been exploited for timber, and secondary forest, with its distinctive flora and structure, has developed along abandoned logging tracks noticeably penetrating logged primary forests. Rising population levels, officially endorsed transmigration programmes as well as spontaneous migration, have also led to the clearing of large areas of primary forest for cultivation. This, in many instances, has subsequently been replaced by secondary forest as often inappropriate farming methods and agricultural schemes have failed. Natural disasters have also added to this proliferation of *belukar*: in 1983, a great fire destroyed three million hectares of mainly logged forest in East Kalimantan: this huge area has now regrown as secondary forest.

SHIFTING CULTIVATION

Secondary forests are closely linked to shifting cultivation (below), the traditional farming system in the rain forest regions of Indonesia. After a fallow period of some seven to ten years, sufficient plant mineral nutrients have accumulated in the trunks, branches, twigs and leaves of secondary forest trees to fertilise one or two harvests. The farmer then cuts the forest, sets fire to the dead brushwood and plants his crops amongst the ash, which provides the source of readily soluble nutrients needed for cultivation.

Rice is the principal crop, but other cereal and vegetable species are also planted, including maize, squash, aubergine, cucumber, chilli pepper and Job's tears — a type of grass grown for its ornamental and edible seeds. A few papaya, bananas and other fruit trees are also often grown. During the second year weed growth becomes increasingly difficult to control and by the end of the second harvest the land is usually exhausted of nutrients and the farm site abandoned again. Fruit trees, however, will continue to provide food for several more years. An experienced farmer can judge when a patch of *belukar* is mature enough to cut down and when it is time to cease cropping. Shifting cultivation is a low–input system, and well–suited to the nutrient–poor soils of the rain forest ecosystem. Contrary to popular opinion, shifting agriculture is sustainable indefinitely; it is only when the bush fallow period is too short, or the cultivation period too long, that the system breaks down.

In the parts of Indonesia where a seasonal climate prevails fire plays a major role in determining the type of vegetation which grows. Where burning is frequent, monsoon forest is replaced by grassland; if fires are infrequent, then shrubby secondary forests develop.

Pioneers and Forest Regeneration

Secondary forest consists of a special group of tree species which are often called 'pioneers', in that they are able to germinate and establish themselves on open sites. These species are well adapted to hot, brightly lit and dry situations, and they generally grow very rapidly. Many have very pale, soft timber and large leaves. Most of these secondary forest tree species only grow to about 20 metres in height, and few live for more than 20 or 30 years. During this time, however, they will have played a vital role in the regeneration of primary forest.

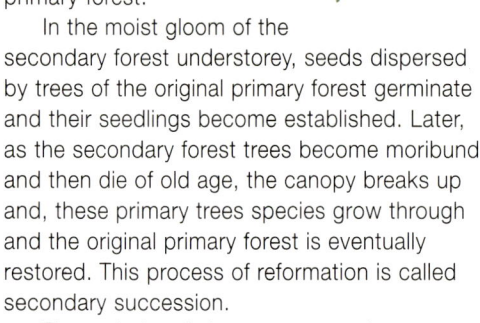

In the moist gloom of the secondary forest understorey, seeds dispersed by trees of the original primary forest germinate and their seedlings become established. Later, as the secondary forest trees become moribund and then die of old age, the canopy breaks up and, these primary trees species grow through and the original primary forest is eventually restored. This process of reformation is called secondary succession.

The majority of pioneer tree species are distributed between relatively few families, with the Euphorbiaceae containing some of the most distinctive genera, notably *Mallotus* and *Macaranga*. Most pioneer species produce copious numbers of small seeds which are easily dispersed by wind or animals over long distances, so that there is a high probability of them reaching suitable growing sites. The seeds of primary forest species, on the other hand, are usually dispersed over short distances, so that the regeneration of primary forest depends on there being a nearby source of seeds. Swidden farmers commonly leave individual primary forest trees, or clumps of trees, standing, so that their seed will invade the bush fallow. Logged forest usually has plenty of primary forest trees left, so that areas of secondary forest are also easily colonised. In contrast, when large areas of primary forest have been cleared, as in the case of abandoned plantations or the recent forest fires in Kalimantan, there may be no sufficiently close seed source available and succession beyond the secondary forest stage is in consequence a very slow process.

Pioneer species have seeds which are capable of dormancy and after their dispersal, accumulate in the soil as a 'soil seed bank'. The bright light and high temperature of a newly formed canopy gap triggers germination from this bank and pioneers growth to fill the gap.

Composition and Structure

Indonesia has several thousand tree species, of which only one or two hundred are pioneers. In this respect, secondary forest has relatively few species per hectare in comparison with primary forest, where there are commonly between 100 and 200 different tree species per hectare. At the same time, the range of different plant forms is more restricted in the secondary forest: woody climbers may be common in some areas, but bole climbers, epiphytes and forest floor herbs are rare. Thus the total number, both trees and all plant species per hectare in secondary forests is but a tiny fraction of that found in primary forest areas.

It is, however, of interest that the weight per unit area of leaves, twigs and small branches is about the same as primary forest, which differs mainly in having a very much greater biomass of trunks and by limbs.

As a result of a great fire in East Kalimantan, in 1983, which destroyed most of the forest, large areas of secondary forest have been generated. Burning coal layers, visible in the foreground, are backed by secondary vegetation.

THE GENUS *MACARANGA*

The composition of any secondary forest in both western and eastern Indonesia will typically include several tree species belonging to the genus *Macaranga*, whose range extends throughout tropical Asia, with a few species also present in Africa. About 30 *Macaranga* species, (*M.gigantifolia*, at left), are found in Sumatra, 50 in Borneo and 80 in New Guinea, though only 16 species occur in the rain forests of Sulawesi and Maluku. A few *Macaranga* species may also be present in primary rain forest undergrowth. The leaves of the latter look very much like those of other rain forest trees, but secondary forest *Macaranga* species have large leaves, which are often lobed, with veins spread like the fingers of a hand. They are easily identified and are often found growing by the roadside; up to ten species may commonly be present at any location.

The fruit of *Macaranga* species are capsulate, often spiny, and split open to expose black seeds, sometimes with a red fleshy jacket (sarcotesta). They are dispersed by birds. As in the case of many other pioneer species, *Macaranga* seeds can remain dormant in the soil for long periods; germination is triggered by the hot, brightly lit conditions associated with a gap in the forest canopy. The natural habitat of *Macaranga*, and other pioneers, are open sites created by landslips and alluvial river banks. Human activities, however, have created huge new areas of suitable habitat and led to the emergence of large forests of *Macaranga* and other pioneers species, which are far greater in extent than would formerly have been the case. Some *Macaranga* species, especially those occurring in western Indonesia, have entered into a symbiotic relationship with ants, mainly *Crematogaster* spp. The ants nest in the hollow twigs of these species and tap their sap with scale insects (*Coccus* spp.). They also feed off starch grains produced by the tree. In return, the ants will attack leaf-eating insects, and nip off invading vines.

The plantain squirrel (Callosciurus notatus) *resting on the branch of a tree, busy feeding.*

PLANT–ANIMAL INTERACTIONS

The bright colours of the yellow flowers attract pollinators such as the bee.

The Malayan tapir (Tapirus indicus) chewing on the tender foliage.

A macaque, a possible seed disperser, feeding amongst the mangrove roots.

Bamboo shoots form part of the diet of the Sumatran orang utan.

Most plants interact with animals at some stage of their lives. The creation of new seed often requires an animal to transfer pollen from one plant to another, and the different strategies employed to achieve this are astounding in their ingenuity. Some plants cheat, however, by luring insects to an apparently luscious treat, only to snare them in a fatal trap.

Many flowers are pollinated by insects so it is not surprising to find a direct correlation between an increase in the number of insects and the onset of the flowering season of the plant species they pollinate. Although some flowers are present in a forest throughout the year, flowering peaks tend to occur during the drier times, and often coincide with, or follow, periods of leaf production which are, in turn, exploited by caterpillars of moths and butterflies. Such patterns of insect abundance also influence the breeding behaviour and population size of insectivorous birds.

Once the fertilised seeds have matured, the assistance of an animal may be sought to move them to a site conducive to germination. A fruiting body may either snag on fur or feathers, or else be eaten and subsequently voided. Not all frugivorous animals are ideal as a means of transport, and sensitive seeds may be destroyed as they pass through digestive tract. Others, however, require this abuse in order to stimulate germination.

Nutritious seeds offer an easily obtainable food source for many animal species. Accordingly, some plants have developed strategies to minimise this premature loss of the next generation. Perhaps the most ingenious ploy is found in the important timber family of dipterocarp trees which have synchronous mass fruiting spasms every five to seven years, it is believed that the sudden superabundance of large oil–rich seeds satiates the animals that eat them, enabling good numbers of seeds to survive to produce seedlings.

Leaves eaten by many animals may seem to be an uncomplicated food source, yet herbivores actually run all manner of risks from the consumption of toxic and digestion–inhibiting chemicals. Most leaf–eaters actually feed on a wide range of plant species, which may help to avoid, or at least alleviate, the worst discomforts of indigestion. Rhinoceros and buffaloes, for example, will consume the tannin–rich leaves of the small tree *Holarrhena antidysenterica*, or those of *Ceriops* mangrove trees, in order to alleviate diarrhoea. Some plants benefit from providing shelter for colonies of aggressive ants which in return attack caterpillars and other leaf predators.

Pollination by Animals

Flowers valued for their beauty abound in everyday life in Indonesia. Grown for decoration around kampung houses, used as offerings in temples and immortalised in batik prints, they do not satisfy the aesthetic tastes of people alone. Flowers contain the sexual organs of plants and the bright colours and heady perfumes serve to attract animals with the aim of using them to assist in the process of reproduction.

*The small, brightly coloured crimson sunbird (*Aethopyga siparaja*), a native to southern Asia, belongs to the family Nectarinidae. Though similar in appearance and habits to the hummingbird, it has a curved bill, and is a true singing bird.*

A bee, a possible pollinating agent of the Arenga *palm, hovers near its flowers.*

The Papilio memnon, *a butterfly feeding on the pollen of the bright orange coloured* Ixora javanica, *an ornamental commonly found in Indonesia.*

The Sex Life of Plants

Plants have far more complex sex lives than most animals. Individual plants can be male, female, or, most often, both. The major problem facing plants when it comes to sex is that they are rooted to the spot. Therefore they must indulge in remote reproduction by liberating packages of male genetic material to fertilise the eggs of other individual plants and hence produce the seed for the next generation of that particular species. These male packets are called pollen grains. They are tiny, dust–like particles which germinate when they settle on the appropriate part of the female structures within a flower. Once this has been achieved, the pollen develops a narrow tube which grows inside the ovary and fertilises the eggs.

The transfer of pollen from one flower to another is left virtually to chance by some plant species. Very large quantities of pollen grains are merely released into the air and carried by the wind. Of these only a tiny proportion of the grains will be successful in landing on a female flower of the right species. Such wind–pollinated species usually have small, inconspicuous flowers because they have no incentive to advertise their presence. The huge and important grass family is largely wind–pollinated, as are some trees such as *casuarinas*.

In the tropical rain forests that cover much of Indonesia, however, there are relatively few species that rely solely on the wind to transfer their pollen. Instead they engage the assistance of animals. Many types of animals are employed by flowering plants as agents of pollination, ranging from tiny insects to birds and mammals. All are involved in a business that varies from obligatory interdependence, through uneasy cooperation to blatant exploitation and treachery. Most plant species specialise in a certain group of pollinators and have evolved a certain flower design that is suited to using that group.

Floral Attractions and Rewards

Animals do not pollinate plants out of an altruistic concern for the well–being of plant populations. They transfer pollen inadvertently while exploiting flowers for certain rewards, or they are duped into visiting the flower by the plant. The incentive offered often consists of nectar, a sugary secretion from glands within the flower. Many insects, particularly bees, butterflies and moths, are reliant on nectar for food, as are a few species of birds and mammals. Pollen may also be collected for food. The rewards the plant has to offer are clearly advertised. Brightly coloured petals, strong smells and a conspicuous position all help to attract the pollinators to the flower. Other plants imitate their pollinators' objects of desire. Some flowers, including the genus *Rafflesia*, mimic rotting carrion to attract flies that feed and lay their eggs on putrefying meat. Certain orchids produce scents that attract male bees or wasps and stimulate them to mate with the complex orchid flower, mistaking it for a female insect. In the process, the flower sticks packets of pollen on the unsuspecting male insect, or accepts such packets if the male has already been hoodwinked into sexual intercourse with another orchid flower.

Insect Pollination

Bees are probably the most widely used group of pollinators. Famous for their industrious ways, worker bees are specialist floral foragers, collecting nectar and often pollen to feed their colony. Bees visit many different types of flowers where their heads or bodies are brushed with pollen. Often possessing long tongues, they can obtain the nectar from deep within long tubular flowers. Some species with relatively short tongues get around the problem by chewing holes in the bases of such flowers. This circumvents the flower's pollen presentation system and hence these species are termed nectar thieves because they do not give any service in return for the floral rewards they get.

Certain species of beetles collect pollen from

flowers and, being rather messy feeders, are good at transferring pollen from flower to flower, though of course each flower loses some pollen as beetle food. Beetle–pollinated flowers are often relatively large with a strong, heady perfume, often slightly alcoholic in aroma. Magnolias and many members of the family Annonaceae are beetle pollinated, as are a number of palms. Certain waterlilies also rely on beetles, and their blooms have been likened to 'beetle nightclubs'. Opening in the evening the waterlily flowers produce a powerful scent to attract the beetles which crawl down into the flower. As the night progresses, the flower closes, trapping the beetles, which (maybe with a little scientific licence) abandon themselves to a frenzy of partying, which includes being liberally coated in lily pollen. Butterflies and moths usually have long tongues, sometimes longer than their bodies, and can exploit very long–tubed flowers. Butterflies, being active during the day, are attracted to brightly coloured flowers, whereas the nocturnal moths are frequent visitors to pale and perfumed blooms.

Birds as Pollinators

Some birds specialise in a diet of nectar. The hummingbirds of the New World are the best known examples, but Indonesia, too, has many species of sunbird which are often as brightly coloured as their American colleagues, but are better acrobats. Unlike hummingbirds, which hover in front of flowers, sunbirds poise delicately on the flowers from which they are feeding. Flowers visited regularly by birds are usually large and red or orange in colour — hues which are not attractive to most pollinating insects.

Pollination by Night

A few species of bat are specialist pollen and nectar eaters, having long tongues to facilitate nectar extraction. A number of well-known trees rely quite heavily on bats for pollination. These include the durian and the *petai*. Bat–pollinated flowers open at night: when bats are active. The flowers are often large and pale and frequently produce a strong, musty odour. Bats, being much bigger than insects, require far greater rewards. The *petai* trees produce large heads of flowers on the ends of long dangling branches. The bats have to bury their faces in the shaving brush–like inflorescence to suck up the nectar and, in so doing, are liberally sprinkled with pollen. A few of the flowers that are successfully pollinated will then go on to produce the big pods which contain the seeds esteemed for sambal throughout the Malay Archipelago.

A small insect visiting the flowers of Sonneratia *sp., commonly found in the mangrove forest.*

The flying fox, the largest bat in the world, belongs to the genus Pteropus. *They are attracted to the well–exposed, bushy yellow flowers of the rose apple, which exhibit features specially adapted for bat pollination.*

Dispersal of Seeds and Fruits

Seeds, like pollen, may be dispersed by wind, water or animals. Dispersal removes seeds from the immediate vicinity of the parent plant, where they are likely to lose out in the competition for essential resources, and places them elsewhere so that they may have a better chance of survival. Although most seeds will end up in unsuitable locations and die, some will find favourable conditions for germination.

Dispersal by wind and water (clockwise): Showing two routes by which the once bare Anak Krakatau, as a result of the 1883 volcanic eruption, was recolonised: light, feathery, wind dispersed propagule floating across the Sunda Straits; Barringtonia fruit washed ashore by prevailing water currents. Dipterocarp fruit, in Kalimantan, with elongated wing–like sepals, adapted for wind dispersal.

The short–nosed fruit bat (Cynopterus sphinx) eating the fruits of jumbu or roseapple.

The greater hornbill (Buceros bicornis) found in Sumatra, is a large bird which feeds on fruits of the fig trees.

Dispersal by Wind and Water

Most seeds contain a food reserve to sustain the early growth of the plant. In this respect, even the smallest seed is relatively heavy when compared to a pollen grain. Wind–borne seeds typically have a modified surface which slows down the rate of fall and increases the chance of being caught by an air current. Characteristic features include fine hairs or papery wings, which may develop from the seed itself or other parts of the flower. The important west Malesian tree family Dipterocarpaceae is distinguished by its characteristic two–to–five winged fruit. Seeds may also be dispersed by water. The fibrous outer coating of the ubiquitous coconut (*Cocos nucifera*) traps air and prevents the penetration of seawater; it is extremely buoyant and has been distributed by ocean currents throughout the tropical world.

Dispersal by Animals

The majority of rain forest plants rely on animals to transport their seeds to new sites. To ensure this, many plants cover their seeds in a nutritious, fleshy fruit. The dispersal mechanism may involve dropping the seed after eating its covering, ingesting and later excreting the seed, transporting them on feet, fur or feathers, or the burying actions of some animals which occasionally fail to retrieve stored foods.

Successful seed and fruit dispersal may depend on a number of anatomical features which have to do with the way a plant's fruit are presented. For example, they may occur individually or be grouped in clusters; they may grow on twigs, branches or the trunk itself. Factors such as these will determine how easily access is gained to the fruit and the type of animal involved in their dispersal. Some fruit will be picked off the tree by arboreal species; others will drop to the forest floor to be eaten by ground–dwelling animals such as elephants, deer, pigs, rodents, cassowaries and even carnivores such as tigers, bears and civet cats. Other important factors include the size, shape, colour, smell, texture and chemistry of the fruit and seeds.

Bats as Dispersal Agents

Bats tend to be attracted to fruits which are a dull yellow or brown colour, and which have a musty smell. These are typically held away from the foliage. In many cases bats, such as the, flying foxes of Indonesia are vital as seed dispersers. They simply suck out the liquid from fruits and discard the seed without ingesting it. Some very small seeds, such as those of figs, may pass through the gut. By contrast, large seeds are rarely swallowed. Bats usually carry the fruit away from the parent tree before eating it. Considerable distances and even the crossing of water may be involved here, and bats, together with hornbills and fruit pigeons, have probably played a critical role in determining the collective range of at least some plant groups found in both east and west Indonesia. The small, weedy tree *Piper aduncum*, for example, is mainly found along roads and tracks which are patrolled by small fruit–seeking bats who favour the fruits.

PLANT–ANIMAL INTERACTIONS

ATTRACTIVE DURIANS
The fruit of the durian tree (*Durio zibanthinus*), far left, has a highly distinctive and pungent odour which is often extremely offensive to those who come across it for the first time. While Alfred Russel Wallace may have believed that it was worth visiting the Far East simply to taste durian, there are many who hold the opinion that this king of fruits tastes like custard passed through a sewer. Disregarding the human perspective, the smell of durian undoubtedly plays a crucial role in its dispersal and is virtually irresistible to rats, squirrels, elephants, pigs, tigers, bears, orang utan (left) and any other animal that can gnaw, kick or claw its way past the ferociously spiny skin. Orang utans will sometimes use a stick to help them gain access to their prize. The odour comprises a complex cocktail of molecules which create a bouquet which is simultaneously reminiscent of fried onions, fruit punch and old socks. The taste is not the only reward of this fruit, since the fruit is high in carbohydrate and fat as well.

Gonocaryum seedling growing from the faeces of the cassowary bird.

Wild figs found on the forest floor, possibly dispersed by species active in the understorey.

Bird-dispersed Fruits
A large number of bird species feed on fruits and there are many different types of fruits which are dispersed by them. Seeds of those primary rain forest trees which have large bird-dispersed seeds tend to be surrounded by a relatively thin layer of nutritious, often lipid-rich flesh. The seeds are ballast and the bird regurgitates or defecates them within a short time. Fruits with small seeds are more likely to have a juicy flesh in which the main nutrients are carbohydrates, often sugars. Many of these fruits are as small as five to ten millimetres, so a wider range of birds are able to feed on them than on the large-seeded fruits. Some of these birds are less dependent on fruits as a source of food than the large fruit-eating birds such as hornbills although it is rare for a bird to feed exclusively on fruits. Bird-dispersed fruits are brightly coloured and may have two or more contrasting colours, such as black and red or white and red, which stand out against the dominant greens and browns of the forest.

Attractive Figs
Fig trees are a characteristic feature of many lowland rain forests and attract animals of every variety when they are in fruit. Some fig species are very large with a pollination strategy that results in individual trees fruiting at different times. As a result, ripe figs are nearly always available in plentiful quantities and, for many animals, they represent a dietary staple. With the exception of green pigeons (*Treron* sp.), which grind up the minute fig seeds, most animals pass the seeds unharmed, after digesting the nutrients from the flesh in which they are cocooned. The regular availability of figs is particularly important to fruit-eating animals because of the seasonal nature of most other fruit species.

Fishy Business
It is likely that fish are responsible for the dispersal of the fruits of some riverine plants but, as yet, virtually nothing is known of this mode in Indonesia. It is said that the common catfish (*Clarias batrachus*), which is found in Bornean peat swamp forests, develops a strange flavour at certain times of the year after it has eaten the aril of the ramin tree (*Gonystylus bancanus*).

Conserving Fragile Links
The long-term conservation of Indonesia's forests depends largely on ensuring that the fragile inter-dependencies of dispersal and the diversity of animal and plant species are maintained. The extinction of a fruit-eating animal in a particular area may lead to the extinction of some plant species there, though this may not happen for many decades or centuries.

The mango (Mangifera sp.) fruits are dispersed by the larger bats.

THE VERY BEST COFFEE
In most Indonesian markets it is possible to find a stall selling coffee beans which, while looking like other beans, are rather more expensive. These beans are called *kopi luak* or civet coffee. They are gathered from piles of faeces left by civet cats on paths in and around coffee gardens. It is believed that the civet (right) has a very discerning palate and picks only the choicest coffee berries which subsequently make the finest coffee beans.

Mistletoes

The mistletoe group is made up of a handful of plant families which share the characteristic of growing parasitically on the branches or roots of other shrubs and trees. Mistletoe plants, however, do bear leaves containing chlorophyll, which enables them to photosynthesize at least some of their own nutrients. Because of this ability they are often referred to as hemi-parasites.

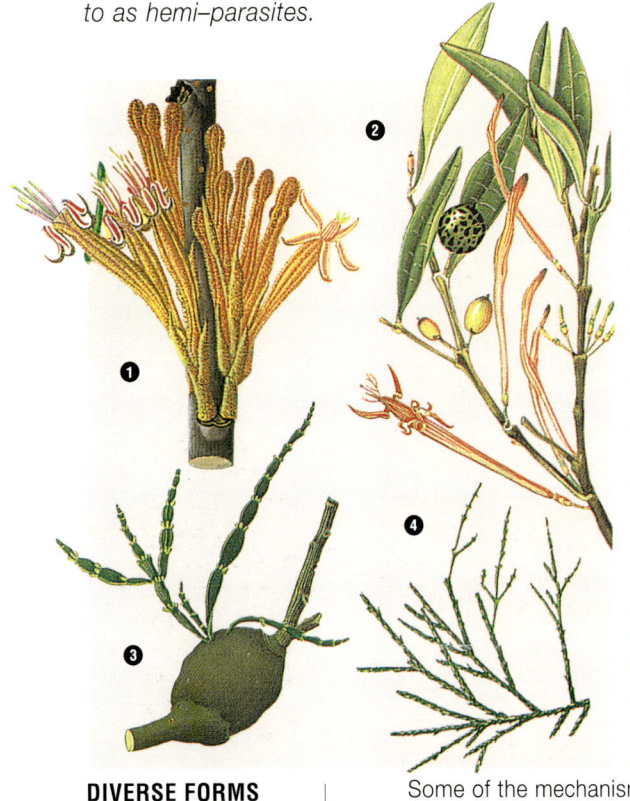

Distribution
By far the largest and best known of the mistletoe families are the Loranthaceae and Viscaeae. Both are found throughout the tropics and are well represented in Indonesia.

Pollination
Apart from the interesting nature of their peculiar parasitic existence, many species are also involved in highly specialised relationships with the birds which visit and pollinate their flowers. Flower peckers or mistletoe birds (*Dicaeum* spp.) are particularly prominent here.

Some of the mechanisms for ensuring pollination are quite spectacular, the most striking feature involves the explosive opening of the flower-buds. Here, the head, or winged part, of the unopened flower is either squeezed or pecked at by the pollinating bird, which causes the corolla lobes to spring apart, in the process, flicking a shower of pollen from the stamens onto the bird.

DIVERSE FORMS AND RANGES OF THE INDONESIAN MISTLETOE
❶ *Dendrophthoe magna*, in West Java
❷ *Macrosolen avenis*, also in West Java
❸ *Korthalsella japonica*, found in both Java and Sumatra
❹ *Korthalsella dacrydii* grows in Java, Kalimantan and Timor

Seed Dispersal
Birds play an important role in the dispersal of seeds of many of the Asian mistletoe species. The soft, inner part of the fruit is swallowed by the bird after it has first removed the tough outer rind. The nutritious part of the fruit is broken down in the specialised digestive tract of the bird and the seed is passed out onto the branch of the host plant in the bird's droppings where it takes root.

The seeds germinate readily and a sucker-like organ is produced. This attaches itself to the branch of the host plant and subsequently penetrates the host's internal tissues. A swelling results at the point of attachment of the mistletoe; this is called the haustorium. A mistletoe may produce one large haustorium or else there may be several small sucker-like haustoria arising from creeping shoots. The larger haustoria are sometimes carved into ornamental figures and are sold locally as souvenirs.

Pests
Mistletoes can propagate very quickly and will eventually overrun their host completely, causing the death of both host and the parasite mistletoe alike. Several of the more common species can become serious pests in fruit orchards and among plantation crops such as rubber trees. The damaging effects of the parasites are well appreciated by orchard and plantation owners who may attempt to weed them out.

Mystical Significance
In many cultures mistletoes are endowed with a mystical significance because of their parasitic habit (the common European mistletoe *Viscum album* is one such example). In Borneo, mistletoes are generally known to the indigenous tribes by the name *benalu*, and preparations of the leaves are employed medicinally to treat headaches and vertigo. It has not been established, yet, whether the plants actually have any beneficial properties.

Hyperparasitism
Some species are hyperparasites, that is, they are parasitic on another parasite. This often involves a Viscaceous plant being hyperparasitic on a member of the family Loranthaceae.

The Bornean *Ginalloa*, a hyperparasite, was found growing on a Loranthaceous mistletoe. *Ginalloa* are closely related to the European mistletoe, *Viscum album*, both belonging to the family Viscaceae. Unlike Loranthaceae, the flowers are small and insignificant, with separate male and female flowers occuring on different plants.

Although brightly coloured, the long–spiked flowers of Helixanthera xestophylla *are small and do not show a high degree of specialisation, as exhibited by other Asian genera of mistletoe such as* Macrosolen.

The long–tube flowers of Macrosolen formosus *are most likely pollinated by spider hunters (*Arachnothera spp.*) with long, slender, curved bills which are able to reach down into the long flower tube. It is possible that the flowers are opened by pecking at the expanded wings of the corolla towards the apex, the corolla lobes then twist and reflex.*

«*Macrosolen beccarii, a forest understorey species exhibiting adaptative features, including an explosive flower–opening mechanism for its pollinator, the flower pecker with its relatively short stout bill.*

THE DISPERSAL AND GERMINATION OF MISTLETOE

A Flower pecker (*Dicaeum* sp.), forms a specific relationship with the mistletoe. They feed on the ripe berries of the mistletoe and void the seeds in droppings. The seeds are coated in a sticky substance called viscin which stick them together in strings and glue them firmly to the branch of a suitable host where they germinate **B**.

C to **G** The seed establishes itself and grows at the host's expense extracting nutrients from the host through the haustorium — a sucker–like organ which appears as a swelling at the point of attachment.

Haustorium

Carnivorous Plants

There are an estimated 550 species of carnivorous plant in the world, of which 66 are found in Indonesia. These plants are found at all altitudes, from sea level to the frosted mountain tops of Irian Jaya, and occur mostly in nutrient–poor substrates which have full exposure to the Sun; very few species are found in the deep shade of lowland rain forest.

Utricularia gibba, *one of the five aquatic species of bladderworts found in Indonesia.*

Utricularia liginosa, a terrestial bladderwort.

Slippery Pitchers

Indonesian carnivorous plants are grouped into four genera, all of which have very different animal–trapping devices. The genus *Nepenthes* is the most conspicuous of these. It includes the world's largest carnivorous plants and is most diverse and numerous in Indonesia. These plants are usually woody climbers, or shrubs, in which the midrib of the leaf blade extends into a grasping tendril which terminates in a lidded pitchers. Such pitchers may reach lengths of 30 centimetres or more. Each is partially filled with a sweet watery liquid which may smell like nectar. Insects, and occasionally small birds and rodents, are attracted to these brightly–coloured plants by the promise of a free meal. Scrambling around the lower lid and pitcher lip to reach the nectar is extremely hazardous on the slippery, loose wax–scales of the pitcher wall. More often than not, the hapless creature tumbles in and, unable to gain a foothold on the vertical, drowns in the liquid. Digestive enzymes reduce the softer parts to a nutrient–rich soup which is then absorbed by the inner wall and translocated around the rest of the plant.

Most species of *Nepenthes* show two quite different types of pitcher on the same plant. The lowermost pitchers are on straight tendrils and face the centre of the plant. They are egg–shaped, with two fringed wings down their front. As the stem climbs, the pitchers abruptly change shape and face away from the centre of the plant. Funnel-shaped, and on coiled tendrils, they lose their fringed wings and catch a different range of prey.

The most common and widespread species of *Nepenthes* occur at the edge of forests or in open swamp forest at low altitude. Examples include *N. rafflesiana*, *N. mirabilis*, *N. gracilis*, and *N. ampullaria*, the latter producing climbing stems that are up to 30 metres long. A greater diversity of *Nepenthes* species, however, is found in high altitude montane forest, where they are particularly common on mossy ridges. Many of these species are rare and, in some instances, may be restricted in range to a single mountain. *N. clipeata*, for example, is known only from Bukit Kelam in Kalimantan and may now be extinct in the wild; other species, such as *N. klossii* from Irian Jaya, have not been recorded since they were first discovered 70 years ago.

THE PITCHER TRAP
Adaptation for the special mode of nutrition as demonstrated by the carnivorous *Nepenthes*, also known as the pitcher plant.

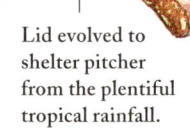

Lid evolved to shelter pitcher from the plentiful tropical rainfall.

Insect investigating the pitcher plant.

Waxy rim which, besides preventing the escape of fallen prey, also secretes the sweet nectar that lures the unsuspecting prey into the trap.

A reserve of digestive fluids found at the base where both the digestion and the absorption of prey takes place.

Bladder Traps

The bladderworts, *Utricularia*, are for the most part inconspicuous plants, yet form the largest genus of carnivorous plants in the world with 214 species recorded, 15 of which occur in Indonesia. Bladder-worts occur in a wide range of habitats: six of these species, *U. bifida*, *U. uliginosa*, *U. baouleensis*, *U. minutissima*, *U. subulata* and *U. caerulea*, are terrestrial plants growing in wet, open areas alongside sundews, sedges, pipeworts and *Burmannia*. Five bladderworts are aquatic — *U. australis*, *U. punctata*, *U. gibba*, *U. minor* and *U. aurea* — some of which are normally found in cold temperate zones, and only occur rarely in Indonesia, at high altitudes, around 3,000 metres. Another three species, *U. striatula*, *U. pulchra* (endemic to Irian Jaya) and *U. steenisii* (endemic to Sumatra), are epiphytes, and grow amongst moss on rocks or tree trunks, normally at altitudes of over 1,000 metres.

Bladderworts lack roots, the flower spikes being

anchored to the substrate by modifications of the stem called rhizoids. The numerous, minute, translucent, bladder-like traps, which give the genus its common name, are usually about two to three millimetres in diameter, although some Australian species may be as large as ten or 12 millimetres across. They are fitted with a trapdoor and only function in moist conditions. The precise workings of the trapdoor are not fully known, but are thought to be as follows. The traps are filled with water and are usually lined with bifid and quadrifid glands. The latter pump water out of the trap, thereby creating a negative pressure inside, causing the side walls to become concave. The prey enters the trap, either by accident or possibly being lured by chemical means (in aquatic species, a funnel of hairs around the trapdoor is believed to guide victims to their doom). When the unfortunate creature comes into contact with the trigger hairs the seal is dislodged and the fluid outside rushes in to equalise the pressure. As this happens, the prey is swept into the trap, the door closes and is resealed. At this point the side walls abruptly become convex in shape. The bifid glands secrete digestive enzymes and the victim is consumed. Traps are reset in as little as 20 minutes.

The prey of terrestrial and epiphytic species is poorly documented, but aquatic species are known to take small fish and amphibian larvae as well as invertebrates. Mosquito larvae are also consumed, and there has been interest in India in using species of *Utricularia* to control malaria.

Sticky Hairs

There are roughly 125 species of sundew (*Drosera*) distributed globally, four of which occur in Indonesia. Sundews have leaf-blades covered with glandular hairs or tentacles which usually take two forms. The short-stalked variety, held close to the upper surface of the leaf, and a longer form. The long-stalked variety is most numerous at the margins and radiates outwards. It is believed that the prey may be attracted by the glistening globules of liquid which surround the spherical, usually red, glands at the tip of each hair. On landing, the prey discovers that these globules are, in fact, a viscous mucilage. In struggling to escape, the prey coats itself with the mucilage which blocks its breathing apparatus and causes it to asphyxiate. The long-stalked glands, on detecting the minute quantities of nitrate in the body of the victim, sweep in towards the centre of the leaf, carrying the prey towards the short-stalked glands. The latter then pump out a fluid containing digestive enzymes which reduce the softer parts of the animal to a nutrient-rich soup. This is absorbed by the short-stalked glands, leaving the husk, or exoskeleton, to blow away. Small flies are the usual prey of sundews, but larger insects such as moths may be consumed by several plants acting together.

*The diverse forms of sundew (*Drosera*) exhibited from left: A red specimen found in the heath forest of Sumatra. Drosera peltata, a perennial, with terminal white-petalled flowers, common in grassy places on the mountains of East Java. Drosera indica, a widespread species of fallow rice paddy fields, an annual known from Sumatra, Java, Sulawesi and Kalimantan.*

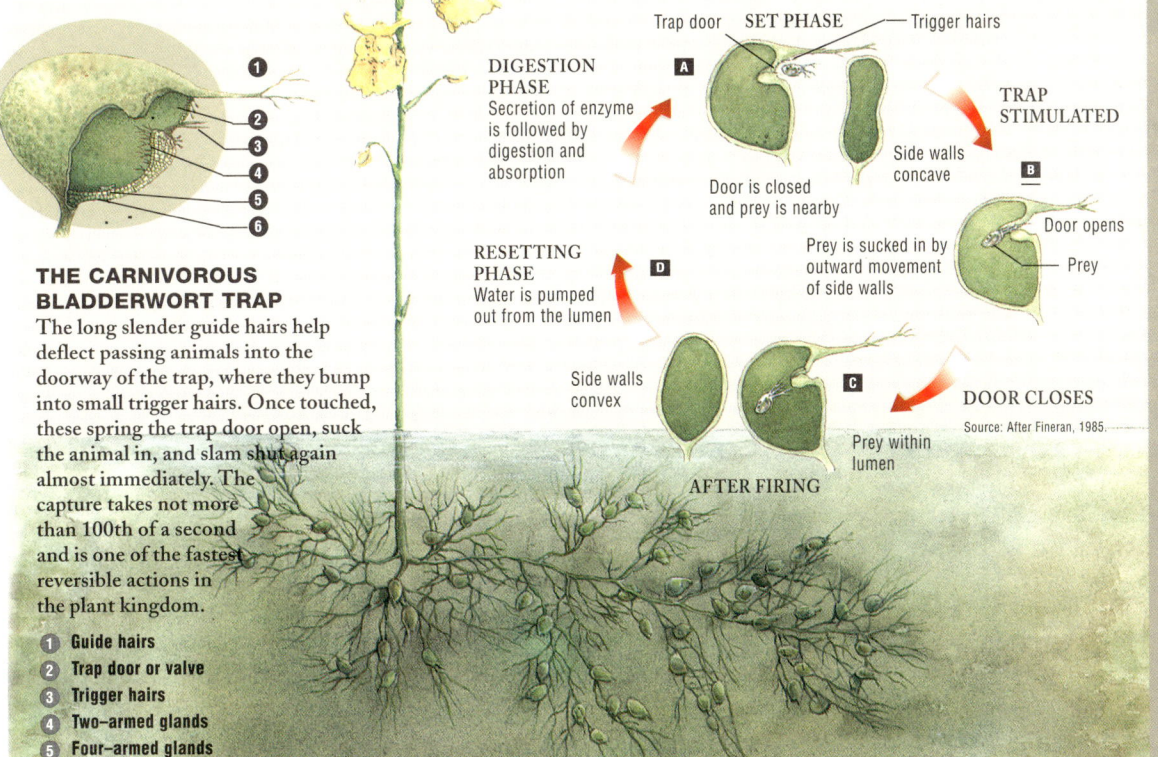

THE CARNIVOROUS BLADDERWORT TRAP

The long slender guide hairs help deflect passing animals into the doorway of the trap, where they bump into small trigger hairs. Once touched, these spring the trap door open, suck the animal in, and slam shut again almost immediately. The capture takes not more than 100th of a second and is one of the fastest reversible actions in the plant kingdom.

1. Guide hairs
2. Trap door or valve
3. Trigger hairs
4. Two-armed glands
5. Four-armed glands
6. Globe-shaped glands

DIGESTION PHASE Secretion of enzyme is followed by digestion and absorption

SET PHASE A — Door is closed and prey is nearby — Trap door — Trigger hairs

TRAP STIMULATED B — Side walls concave — Door opens — Prey is sucked in by outward movement of side walls — Prey

DOOR CLOSES C — Prey within lumen

AFTER FIRING — Side walls convex

RESETTING PHASE D — Water is pumped out from the lumen

Source: After Fineran, 1985.

MECHANISM OF THE BLADDERWORT TRAP

A A vacuum is created when water is pumped out from the trap by special hairs on the bladder wall.

B The trap door is triggered opened by the movement of prey near the entrance, both water and prey are sucked in.

C The prey is trapped with the immediate closure of the trapdoor.

D The plant then digests the prey with the help of secreted enzymes.

Plant Defences Against Herbivores

A plant needs its leaves to generate energy by means of photosynthesis, and to circulate water and nutrients through transpiration. Obviously they cannot afford to lose too many, too often, to herbivorous animals, so many plant species have evolved elaborate defence systems to protect themselves from being eaten.

Leaf edge of palm, Metroxylon sp. with protective spines which acts as a deterrent to the herbivores.

»The equally nasty spines exhibited by the salak or snakefruit palm (Salacca edulis).

↗Herbivores feeding on the undergrowth and leaves of young seedlings (from top): Asian elephant (Elephas maximus), and the Sumatran Rhinoceros (Dicerorhinus sumatrensis).

The Vogelkop tree kangaroo (Dendrolagus ursinus), one of the two kangaroos endemic to Indonesia, is found in Cenderawasih and the Fak Fak area.

Types of Defence

Plants defend themselves in many ways, the most appropriate form of defence largely being determined by the type of herbivore most likely to attack the plant. Other important factors include the need to ensure that the energetic cost of defence to the plant is less than the cost of replacing vegetation lost to herbivores, which in turn depends on the plant's location and lifestyle.

The final defence strategy may therefore depend, for example, on whether the plant is growing on good soil where nutrients are readily available, or on poor soil where nutrient replacement is difficult. Other factors which may be important include whether the plant is a fast-growing coloniser species or a slow-growing climax forest tree.

Defence Systems

Plant defence systems may be either mechanical or chemical in nature. Thorns are a common feature of tropical rain forests, especially on the trunks and stems of palm species. Most defences, however, are chemical. The two main types of chemical defence are toxins, which poison the animal eating them, and digestion inhibitors, which prevent the animal from digesting its food and obtaining nutrition from it. All plants have both toxins and digestion inhibitors, but the amount and location of each varies between species, their lifestyles and habitat. Plants can also change the chemical composition of their leaves rapidly, and to order. They can, for example, produce more defence compounds when a herbivore starts to attack them. Further, the levels of different chemical compounds in a leaf can vary according to where it is located on the tree, or even with the time of day or night.

Toxic Products

Toxins are generally small chemical molecules and are therefore cheaply produced in terms of energy consumption. They poison the animal which eats them in a variety of ways. They may, for example, interfere with the nervous system, respiration, or the ability of the blood to transport oxygen. Common plant toxins include alkaloids such as cyanide and strychnine. Since toxins are easily synthesised and can be produced relatively quickly, they are often found in fast-growing, colonising vegetation.

Inhibitors to Digestion

Unlike toxins, digestion inhibitors are large molecules and are energetically expensive to produce. They are most commonly found in trees in climax forests. They are often integral to the structure of the plant and act by inhibiting digestion in the animal which eats them. The most common way of achieving this is by either making the material difficult to break down, or by binding onto the animal's digestive enzymes, and in this manner, reducing their efficiency. Some digestion inhibitors act in both ways. Lignin, which forms part of the structure of plant cell walls, as well as tannin, whose action on animals is ambiguous, are among the most commonly used digestion inhibitors.

Another function of digestion inhibitors is that they may prevent unripe fruit from being eaten before they are ready for dispersal. Raw bananas, for example, owe their dry taste to the presence of such compounds.

How do Animals Deal with Plant Defences?

Plants and animals do not follow separate, insular evolutionary paths, but rather develop together in a process called co-evolution. If a plant evolves a simple defence mechanism, then an animal species is sure to develop a way to counteract it. When this happens, the plant will, in turn, have to devise a more sophisticated way of defending itself. Thus, at any one moment in time, what we see is simply the latest stage in an ongoing evolutionary process.

In the rain forest canopy, the main leaf-eaters are insects and some species of primate; vegetation closer to the ground may also be eaten by ungulates such as tapir and deer. Animals have evolved many ways of dealing with plant chemical defences, the solution being determined in part by the animal's size, their mobility and physiology. A caterpillar, for example, has a fairly restricted mobility and as it cannot easily go from tree to tree, therefore favours a different strategy to that of a monkey or deer which can move from one tree species to another, and which are therefore able to choose and feed from a wide range of plant species. No animals can deactivate digestion inhibitors since, by definition, they cannot easily be broken down. Large-bodied species, however, with lower metabolic rates, can eat foods which are less digestible to smaller animals, and can generally cope with higher levels of such digestion inhibitors. Toxins, on the other hand, are smaller, simpler molecules, which can be deactivated by some herbivorous species. This ability is only present in some animals; gibbons, for example, are unable to break down alkaloids and must avoid eating leaves with an alkaloid-based defence mechanism. Other species, however, have evolved ways to cope with the problem, thereby allowing them to eat foods not available to other species.

Some invertebrates have developed enzymes to break down specific toxins, but must in consequence spend their whole lives feeding on only one or two species of leaves. Large-bodied vertebrates, on the other hand, cannot be restricted to such a limited choice and must be able to cope with a wide range of toxins occurring in diverse plant species.

AN INVERTEBRATE'S SOLUTION

Certain members of the butterfly family Danaidae, the milkweed butterflies (right), have solved the problem of plant poisons in a unique and highly specialised way. The caterpillars are voracious leaf-eaters and possess a sac off their gut into which leaf toxins can be channelled before they can cause harm. The sac remains intact, even when the caterpillar metamorphoses into a butterfly, and subsequently provides the butterfly with its own defence system: the poison in the sac is still toxic, and would harm any creature which preyed upon the butterfly. In this way, a toxin designed to be harmful to insects ultimately ends up defending them!

A VERTEBRATE'S SOLUTION

Colobine monkeys in Indonesia include the leaf monkey and proboscis monkey (right). It has a characteristically comical, large, long and pendulous red nose and a distended-looking belly. This is normal in an animal having to process masses of vegetation at a given time.

These monkeys have complex sac-like stomachs. The first, greatly enlarged chamber contains a huge array of bacteria. These deactivate poisons ingested in the diet before they enter the bloodstream, thereby allowing the monkey to eat toxic foods which would kill any primate with a simple stomach.

«*Although small in size, invertebrates can eat huge amounts of plant material because of their large numbers.*

A pig-tailed macaque in the canopy layer, searching for young leaves to feed on.

Plants and Ants

Indonesia has a remarkable number of plant species which have developed close and mutually beneficial associations with ants. In many instances, a plant will provide ants with food resources and nest-sites, while ants, for their part, will either protect the plant from herbivores or supply their host with nitrogen and phosphates. Symbiotic relationships of this nature between two higher organisms is relatively rare and, in the case of plants, only occurs with ants.

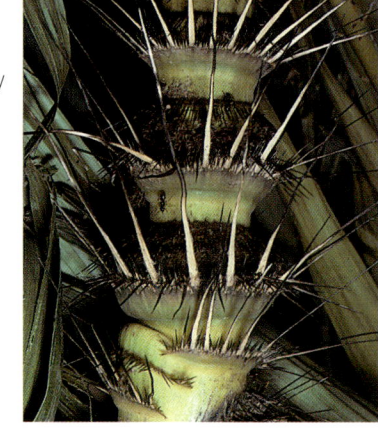

»»*Spine galleries on rattan — Daemonorops sabut act as covered tunnels in which ants live. They are formed as a result of interlocking of spines which are united at their base.*

Section of the ellipsoid shaped leaves of the Dischidia *sp. revealing a mass of branching roots where the ants live. The ants provide nitrogenous matter from their excreta and also bring soil to these roots.*

Tailor ants, common in Asian forests, build their nest by folding and pinning a suitable leaf with their mandibles and feet. The leaf is then stitched together using the silk of a larva.

Protection by Ants

The role of the ant as protector typically occurs with secondary forest plant species and is especially associated with trees that have nectaries on their leaves. These extra-floral nectaries are visited by ants which feed on the sugary solution they secrete. In return, the ants will attack any other creature that they come into contact with while feeding on the leaf. This includes, not only other insects, but also larger herbivores. These bodyguards of mercenary ants are most commonly found on leguminous tree species in secondary forests, where the conditions for rapid growth make the sugary nectar an easy substance to produce and the risk of serious leaf loss or damage by herbivorous insects is considerable.

Ant-Trees

Several species of tree in Indonesia have hollow stems which are inhabited by ants. *Macaranga* and *Endospermum* (Euphorbiaceae) both have pithy stems which are holowed out and inhabited by species of *Crematogaster* ants. *Macaranga* provides remarkable glycogen-rich food bodies, called Beccarian bodies, on the underside of the stipules. Glycogen is a storage carbohydrate normally found only in animals and the ants feed these food bodies to their larvae. It has been shown that in return the ants protect the trees from other insect pests.

Epiphytes and Other Plants

The most diverse and bizarre of all ant-inhabited plants are the epiphytic varieties. Typically, the plant provides a nest-site in return for nutrients. In some species the ant-plant relationship is fairly non-specific and requires no special adaptation on the part of the plant. In others, however, highly specialised and complex structures have evolved which cater for very specific ant inhabitants. In the case of the bird's nest fern (*Asplenium nidus*), the numerous over-lapping leaves form a basket which traps detritus falling from the tree canopy. Many invertebrates live in these baskets and the decomposition of organic matter releases nutrients to the fern.

Dischidia is a genus of climbing epiphytes with several species exhibiting two distinct types of leaves: one of these is perfectly normal, but the other is modified to form an inflated sac, providing a nest for ants of the genus *Phylidris*. Ants use old leaves as middens and adventitious roots proliferate in the rotting debris allowing the host plant to absorb nutrients such as phosphates and nitrates. It seems likely that the plant also benefits from carbon dioxide released by the ants and the debris because the inner surface of the leaf sacs has many stomata. Many understorey rattans have formed special relationships with ants. In the genus *Daemonorops* ants inhabit tunnels between spiny flanges of the leaf sheaths. Other rattans also have ant-inhabited structures, for example the inflated ochrea of Korthalsia species. It is said that when disturbed by a herbivore, such as an elephant, the ants shake the stem in unison, the noise being a warning of an impending attack.

Ant-ferns (*Lecanopteris*) also have striking adaptations similar to house ants. The rhizomes of these plants are swollen and hollow, sometimes with a system of complex side chambers off a longitudinal gallery. Ants live inside these chambers. Experiments have shown that the fern makes use of the nutrients released by the ants.

Ant-House Plants

This group has large hollow tubers which provide a home for colonies of ants. Derived from the hypocotyl, they consist of a series of cavities. Some of these cavities, or chambers, are smooth walled, while others are lined with rough warts. The latter are almost certainly absorptive structures: fungi flourish where the ants defecate and the plant absorbs a range of nutrients from these materials.

There are four related genera of ant-plants in Indonesia: *Hydnophytum*,

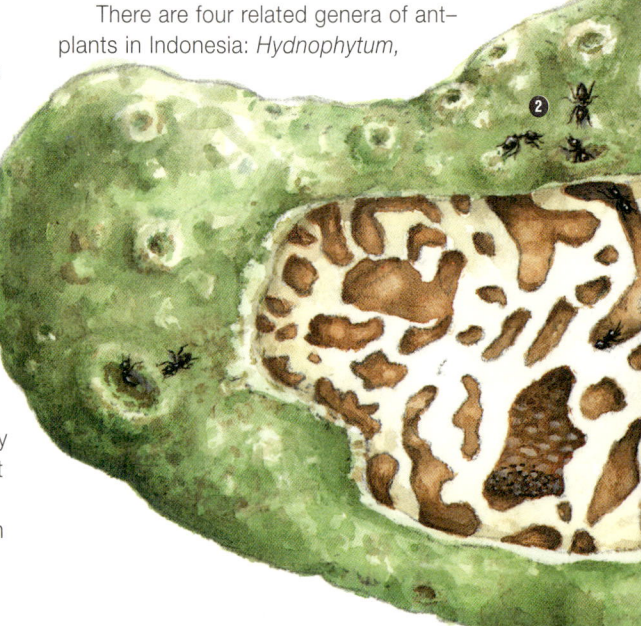

Hydnophytum – A COMMON RUBIACEOUS MYRMECOPHYTE

Hydnophytum belong to a group of epiphytic ant-plants, known as myrmecophytes. They are well developed in forest on nutrient-poor soils, for example heath forest, peatswamp and montane forest, but are also found in all forest types. In this true symbiotic association, the host houses the ants in a series of large chambers in its modified stem. Some chambers also produce fruit bodies. In return, the ants provides a source of mineral nutrient for the plant.

❶ The hollowed stem manifests itself as a tuberous structure.
❷ Ants (*Phylidris cordatus*), at one of the entry points.
❸ Section demonstrating the hollow cavities with its inhabitants carrying out their daily activities.

MACARANGA

The young stems of this secondary forest euphorb are hollow. They are inhabited by ant species of the genus Crematogaster who chew their way through the epidermis to reach the hollow centre. Special food-bodies, called Beccarian bodies, are formed on the underside of the recurved stipules. Named after the Italian botanist, Odoardo Beccari, these Beccarian bodies are rich in protein and lipids. The ants collect them and feed them to their larvae. Some species of other genera also have ant-inhabited hollow stems; for instance Endospemum and Clerodendrum.

Myrmephytum, *Myrmecodia* all Rubiaceae. *Hydnophytum* is the most abundant, with *H. formicarum* being found throughout Sumatra, Java and Borneo, the last also being home to *H. perangustum*, noted for its long, narrow leaves. A wide variety of *Hydnophytum* species are also found in Irian Jaya, including the lowland *H. petiolatum* and the common coastal *H. moseleyanum*, characterised by fleshy, often lozenge-shaped, pale green leaves. There are also a number of local montane species. In the latter, the hollow tubers usually contain water in which a variety of non-specific invertebrates live.

The genus *Myrmephytum*, found in Sulawesi and Irian Jaya, is distinguished from the other genera by having six, rather than four, petal lobes. *Myrmecodia* species, the most specialised of the ant-house plants, are distinguished by their abundant spines covering the tuber and stem. These provide protection for the ants moving about on the surface of the plants. They have walled chambers arranged like shelves and it is here that the ants deposit their brood. The warted chambers are narrow and tunnel-like and, as in the case of *Hydnophytum* species, this is where the ants defecate and deposit other waste materials. Fungi grow here, breaking down the organic debris and enabling the plant to absorb nitrogen- and phosphate-rich substances. *Myrmecodia tuberosa* is the most abundant species of its genus; it occurs at low altitudes from Sumatra to Irian Jaya. It is inhabited specifically by the ant *Phylidris cordatus*, which is also found in other lowland epiphytic ant plants such as *Lecanopteris sinuosa* and *Dischidia rafflesiana*. These plants form a characteristic community and are often accompanied by *Dischidia nummularia* and the orchid *Acriopsis javanica* both of which are typically associated with these ants, but do not appear to offer them a reward.

A wide variety of ant-house plants are found in Irian Jaya. Of these, *Myrmecodia platytyrea* has shields at the base of each petiole, covering the stem along which the ants move to collect nectar from the flowers, while being well protected from birds and lizard predators. On areas of nutrient-poor soils and at high altitudes, species of all three genera grow on the ground rather than epiphytically. Around the Angi Gigi lakes, *H. cecilia* has upward pointing entrances to the cavaties which help the plant to trap rainwater.

The Macaranga hypoleuca, *a common species found in disturbed and well-lit sites:*
A Ants feeding on white food-bodies.
B Hollow, ant-inhabited twigs.

Myrmephytum arfakianum *an epiphytic antplant which also occurs at ground level on the nutrient-poor soil of Irian Jaya.*

*A boy (left) feeding on the sago pith after the ceremonial cutting of the sago palm (*Metroxylon sagu*). This is a staple of the Asmat of Irian Jaya.*

The sale (right) of forest produce, the bright red pandanus fruit, by Dani men in Wamena market in the Baliem Valley, Irian Jaya.

Dayak (below) bundling harvested rattans collected from the rain forest. These were collected near the Ella Ulla River in West Kalimantan, which provides easy transport back to the village.

PEOPLE AND PLANTS: FOREST PRODUCE

The forests of Indonesia have, since time immemorial, provided people with a vast array of economically useful plants. Apart from a range of edible roots, beans, leafy vegetables, fruit, sugars and starches, herbs, infusions and alcohol; plants have been used as building materials, weapons, tools, water carriers, baskets and other household items, fibres, jewellery and ornaments, cloth, musical instruments, medicines, preservatives, pesticides and poisons, oils and resins, soaps and shampoo. Some of these products have considerable economic value which is now widely recognised; others have become less important or been superseded by modern alternatives.

The timber industry is the most lucrative and immediately obvious consumer of Indonesia's botanical resources. The export of timber and plywood totals some ten million cubic metres per annum equivalent to a cube of wood with sides twice the length of a football pitch! Indonesia's timber resources have generated a revenue worth millions of dollars to the State, while creating thousands of jobs for Indonesia's work force. The environmental cost, however, has been high. To date, the timber industry has used a relatively small number of native species, but as the nation's forests become increasingly depleted, and as timber processing techniques become more sophisticated, it seems likely that a wider range of species will be exploited.

Apart from the timber industry, there is also a considerable trade in other forest products such as rattans, resins, medicinal plants and bamboo. Bamboo — a type of grass — is generally gathered from the wild. It is used extensively in Indonesia for building materials, scaffolding, water carriers and conduits, cooking tubes, basketry, mats, weapons, tools, fishing tackle, musical instruments and as a source of pulp for paper. The young shoots may also be cooked as a delicious vegetable.

Most ethnic groups in Indonesia have traditionally employed medicinal herbs and other plant products in the treatment of disease. These cures continue to be used widely even though modern, Western medicines are now easily obtainable. Most remedies are drunk as general tonics rather than as cures for specific ailments. Some traditional recipes have proven therapeutic value but others are no more effective than a placebo. A considerable proportion could be classified as poisons, but in small quantities these are probably harmless. Some countries, however, have banned the import of Indonesian herbal medicines because of alleged carcinogenic effects.

Food from the Forest

Indonesia's forests play an important role in regulating global carbon, nitrogen, water and energy flows; their timber is a source of considerable wealth for entrepreneurs; and they are home to a considerable proportion of the world's animal and plant species. Indonesians use more than 6,000 species of flowering plants. Hundreds of species of fruits, nuts, roots and seeds are obtained from lowland forests, which are also a rich source of wildlife (mammals, birds, fish and insects) and secondary edible products such as wild honey.

Smoking, a process involved in the collection of honey from the forest, in Central Sumatra.

Buckets of sago ready for sale in a local market in Maluku.

Mature rambutan (Nephelium lappaceum), a red and hairy fruit with a sweet taste, indigenous to Indonesia.

A Dayak with prepared bamboo shoots for cooking. The shoot cover has been removed and washed in the Ella Ullu River in West Kalimantan.

Bountiful Produce

The social and economic value of natural products from the forest has long been recognised: trade in forest products has a long tradition in Southeast Asia with various tribes exchanging rattan, resins, camphor and *gutta percha* for salt, iron and other essential materials.

Many non-timber forest products are still of considerable value to local and national economies in tropical countries. In Indonesia, receipts of more than $200 million have been recorded from the collection and export of rattans in a single year. While commodities such as rattan are traded openly in major markets, others such as wild plant foods — fruit, root crops, tubers and medicinal plants — are also actively collected and find their way to local and occasionally national markets. Few, however, reach the international market place and for this reason the true economic value of these foods is often overlooked. In reality, these products are probably worth far more than all timber products combined.

A recent census showed that at least 940 species of plants are used in traditional medicines in Indonesia. Some 450 species of fruit and nut producing plants have also been identified; about 100 species of cereals and root crops provide vital carbohydrates, while a similar number of legumes ensure a supply of minerals and vitamins to local communities. Indonesia is home to a diverse range of fruit plant families including the Alangiaceae, Bombacaceae, Guttiferae, Dipterocarpaceac, Palmae, Euphorbiaceae, Leguminosae, Moraceae, Sapindaceae and Tiliaceae. Many of these are better known by their common names (durian, *Durio carinatus*, illepe nut, *Shorea macrantha*), but the range is extensive and includes wild bananas, rambutan, mangosteen, langsat, jackfruit, sugar palm fruit and other species, many of which are indigenous to these islands. The diversity of certain plant groups is also impressive: at least six different types of the matoa fruit (*Pometia pinnata*) have been recorded from Irian Jaya, some 300 mango varieties have so far been identified, and 19 of the 27 wild durian varieties found in Malesia are found in Kalimantan.

Fruits in all Seasons

One of the underlying factors which explains the abundant and varied wildlife of Indonesia's rain forests is the range of fruiting trees which grow here, producing a wide range of edible food throughout the year. Trees differ considerably in the type and quantity of fruit they bear; some such as the durian have large fleshy fruit while others produce tiny fruit bodies that are crammed full of seeds. Many trees produce fruit once a year and at the same time each year, but some species such as figs may stagger the timing of ripening of fruit: the same species of fig may even bear fruit at different times of the year, a feature which may be controlled by local soil or microclimatic variations.

Although fruits are therefore widely available, their supply is at times irregular and occasionally not substantial enough in dietary terms to constitute a staple food source for, both, large herbivores or frugivores. When certain fruits such as the durian are ripe, however, many of the forest's largest species elephants, rhinoceros, tapirs and even tigers — come from long distances to feast on the rich pickings. Most animal species, however, are forced to diversify as they would be extremely vulnerable to periods of fruit shortages and could easily face starvation.

People also face this dilemma as few fruits can be stored for any length of time — the durian being an exception as this fruit can be cooked into a concentrated paste which can be kept for a long period without spoiling. For humans therefore, fruits are most often seen as a highly valued source of variety and relish in the diet rather than a staple.

Basic Ingredients

Local people everywhere take advantage of wild foods. On the island of Siberut, for example, people rely on sago (*Metroxylon sago*) for their staple food, while in the absence of cultivated sources of carbohydrate the Kubu people of southern Sumatra find their basic starch staple in the underground tubers of a number of climbing yams belonging to the genus *Dioscorea*. Scientists have identified 17 species of yams in Sumatra, but the Kubu have a different system of classification: they recognise about a dozen varieties which are referred to as *benor*, and an equal number of 'other' varieties. These are often told apart by determining whether their stems twist to the right or left, and by the presence of simple or compound leaves.

Benor tubers are regarded as the Kubu's main source of food within the forest and require little cooking. These yams may be detected and identified by their very small scented flowers. The tubers are deeply buried so require greater physical labour to dig them out. Other tubers growing much closer to the surface, are therefore, more easily obtained. But some of these contain toxic alkaloids and tannins which are primarily aimed at deterring herbivores. Most of these tubers can be rendered safe for eating by either soaking them for up to a week or by baking them in the embers of a fire. One species of yam, *gedung* (*D. hispida*) is so poisonous that a piece of raw tuber the size of an apple is enough to kill a human; thorough cleansing is therefore required before this tuber can be safely eaten. Despite this trait, it is a popular species because it can be propagated vegetatively in swidden gardens, and even in primary forests, and constitutes an important and traditional source of food in Indonesia at times of famine. Another species, *tuba hubi* (*D. piscatorum*) has high concentrations of saponins and is widely used during the dry season as a fish poison.

Although the Kubu, like many other forest-dwelling groups, derive their dietary carbohydrate from a variety or tubers, sago-yielding palms, fruit and honey, they find that food gathering in the tropical rain forest does not make for an easy living. As a result, they generally prefer to practice swidden cultivation or obtain food in exchange for forest produce.

The Importance of Conserving Species

Tropical forests are a natural storehouse of different species and the importance of fruit trees is widely recognised. Many seasonal fruit trees are planted in swiddens where they are tended to by the removal of competing vegetation, even long after the field may have been abandoned for agriculture. Such trees and their fruit are widely recognised as the property of those who planted them, or their heirs. This practice plays an important role in conserving the genes of certain species. In many forest regions, local communities have maintained and encouraged species diversity by cultivating seed collected from fruit trees, particularly mango, durian and rambutan. One species of mango (*Mangifera casturi*) is now found only in village gardens in south Kalimantan as its natural habitat has been destroyed through forest clearance for agriculture.

Transporting pandanus fruits, collected from the forest, back to the village. The bright red fruit is a supplement to the diet of the Dani people, whose staple consist mainly of sweet potatoes grown in their ditched gardens.

«*Collecting paku, an edible fern, from the forest undergrowth in Tanah Merah in East Kalimantan.*

A Kubu family snacking on palm cabbage: Palm buds or 'cabbages' are easily collected but are not very satisfying and are used only to complement meat. The latter is unusually abundant especially during the mass migration of bearded pigs (Sus barbatus).

Women and girls undertake the major task of digging for benor. The hard work of foraging is compensated for by the immediate edibility of these tubers.

The processing of gedung by a river: Gedung, a species of yam, is highly toxic and requires elaborate processing before it is safe to be eaten. The tuber needs grating, neutralising and washing for several days. When dried, the resulting meal can be stored and easily transported.

The Austronesian Botanic Idiom

For Indonesians throughout the Archipelago, plants have ritual as well as utilitarian uses. The early Austronesians were cultivators who brought with them an intimate knowledge of specific plants and cultigens. Rice and millet, and probably Jobs' tears, were among these early cultigens. As they spread through the islands, the Austronesians adapted their knowledge to new environments and developed further forms of cultivation and plant use.

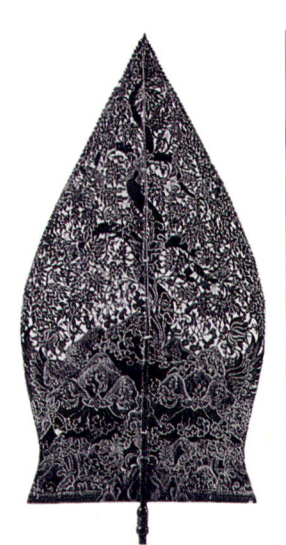

The Javanese Tree of Life, transformed into a mountain or gunung, features as a key image in the shadow puppet theatre (wayang kulit).

Right, clockwise: A harvest offering dedicated to Dewi Sri, is placed in the fields during harvest time. Dewa Kaki, or Dunding, who has enormous sex organs is another important character in the fertility rite. During the dewa sanga, a yearly feast held in Selat for the standing crop, the marriage between Dunding and Dewi Sri is celebrated with huge offerings. Fields with newly planted rice and an altar for offering. An example of a bamboo offering pole among rice fields

Botanical Metaphor

In the course of their long history, the different Austronesian-speaking peoples elaborated many of their most important rituals around the use of plants. Ideas about life and well-being, about derivation from a common source and even about renewal after death came to be phrased, by analogy, with the growth and multiplication of plants.

Common therefore to different Indonesian populations is a similar botanic idiom for social as well as ritual purposes. This idiom, and much of the associated ritual, focuses on key plants that serve as botanic 'icons'. These plants are among the oldest cultigens of the Austronesians: rice and millet, coconut and areca nut, yam and taro, sugarcane and banana. Bamboo is also another important icon in rituals as is, in more general terms, both the dedap and the waringin trees.

A good example of the use of a botanic idiom is the widespread concern to trace and recount the origin of persons, objects or practices back to initial 'trunk' or foundation. Many groups hold origin feasts to honour their ancestors, or celebrations to retrace the initiation of an important activity. Among the

TREE OF LIFE
Among the Ngaju Dayak, the myth of creation also centres on an immense tree. This tree is described as the tree of gold and ivory:
The tree of life puts forth leaves of gold
It bears tender leaves of gold
The tree of life bears blossoms of magnificent gold
And fruit shining like white ivory.

A quarrel between two birds, one male and the other female, atop this primal tree destroys the tree and gives rise to the existing world. In a similar idiom, the Rotinese of eastern Indonesia describe the world metaphorically as a great banyan tree. One ritual chant states:
The banyan has eight branches and the waringin has nine boughs
Nine roads indeed and eight paths as well.
The road of well-being is there and the path of death is there.
Therefore watch with care and look with attention.

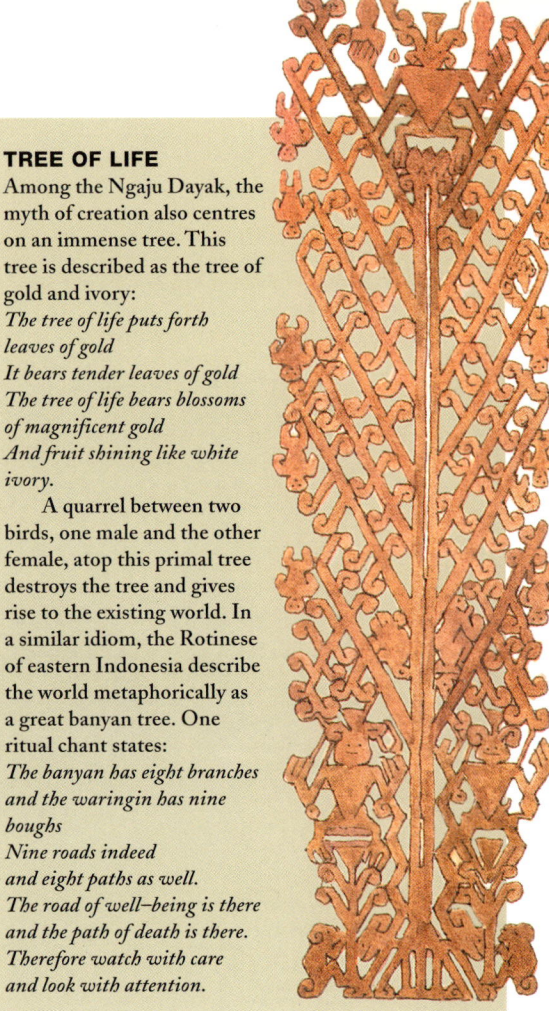

Javanese and Balinese, such origins are referred to as *kawitan*, from the term, *wit*, 'stem, trunk, or tree'; among the various people of Timor, a similar 'trunk' metaphor is used to refer to ancestral origins: for the Atoni Pah Meto, the term is called *uf,* for the Mambai, it is *fun;* and the Tetun, it is *hun.* All of these terms on Timor derive from the same proto-Malayo-Polynesian word, *puqun*, meaning 'tree, trunk, base'.

The Toraja refers to their origins with the term, *rapu,* which also means 'trunk' or 'base'. A family group may be described as a *rapu tallang*: a 'cluster of bamboo' with many stems sprouting from a single clump. On the island of Savu, other botanic metaphors are used to describe social groups. On Savu, maternal moieties are referred to as the Greater or Lesser 'flowerstalk' (*hubi*) and the groups within these moieties are called 'seeds' (*wini*).

The pervasive capacity to speak about all aspects of life in terms of botanic imagery is characteristic of this botanic idiom.

Rice Cultivation

Perhaps the most significant rituals involving the personification of plants are those associated with the cultivation of rice. Although there are a variety of myths about the origin of rice, the most common and widespread tells of the death of a young girl from whose body the first rice is derived. This myth is associated with Dewi Sri in Central Java, Lise, among the Toraja or Nyi Pohaci in West Java. A common practice was to mark out a central section in the rice field in which the first ceremonial rice was planted. This clump of rice also became the first rice to be harvested and would then be treated as if it were a delicate maiden child that had to be gently conveyed, with all the proper ritual, to the rice barn for storage and replanting in the following year. Since rice planted at the centre of a field is the least likely to be pollinated by rice from other fields. This ritual practice of maintaining each field's original rice contributed, over time, to the perpetuation of distinctive strains of rice.

Marriage and Fertility

Using this botanic idiom in ceremonies, just as plants can be personified, by the same token, persons can be spoken of as plants. In asking for a girl's hand in marriage, some groups make their request for new 'seed'. Marriage proposals are often signalled by the offer of areca nut (or a combination of areca and betel). The chewing of betel and areca nut is culturally interpreted as symbolic of the union of sexes.

Like the areca nut, the coconut is often identified as having cooling, female and potentially fertile qualities. In some societies, the coconut is used in performances of the marriage ceremonies. Among the Rotinese, the ritual of marriage features cutting of a coconut with the following pronouncement:

This coconut has five layers
The husk embraces the shell
The shell embraces the flesh
And the water embraces the germ
So let it be
For this young man and woman
Let one embrace the other
And one cling to the other
So that the shoot of the coconut may grow

And the germ of the areca nut may sprout
So that she may give birth to
nine times nine children
And that she may bring forth
eight times eight children.

In keeping with this idiom of life, myths of a primal tree are common throughout Indonesia. In the creation myth of the Tetun, which describes the beginning of the world as the emergence of a huge banyan, boys are the 'fruit' and, girls, the flowers of this tree.

Plants in Rituals

Like many other Indonesian people, the Karo Batak of Sumatra have traditional ceremonies for calling the spirit of rice to enter their fields and provide a good harvest. They also have ceremonies to coax a spirit back to someone who has fallen ill or to entice a spirit to enter the womb of a childless woman. These ceremonies involve the manipulation of numerous plants, and the accompanying chant uses many plant metaphors, some of which, as the spirit is about to enter the woman's body, verge on the erotic:

A bunch of bananas,
There is one in the beginning,
There will be many in the end,
The fruit is bitter in the beginning,
It is sharp in the middle,
But it is sweet at the end.

At a Javanese wedding, an auspicious bundle of plants, known as *tuwuhan*, representing the riches of nature, is used to decorate the entrances of wedding halls. It combines the major ceremonial plants of the Austronesians: rice, sugarcane, banana and coconut with leaves from the *beringin* and the *dedap*. It is symbolically one of the most striking composite botanic icons in the Indonesian world.

The flowers and developing fruit of the banana (Musa sp.). *The flowers are shed as the fruits start to develop.*

In Bali, the areca nut — seed of the areca palm (Areca catechu), *betel leaf* (Piper betle) *and lime paste are prepared as an offering.*

Mangosteen–shaped containers from a betel set. Below: A Balinese offering of champak (Michelia champaca) *and sandat* (Cananga odorata) *flowers.*

TUMPENG ROBIYOK

At the beginning of the Islamic calendar year, I Soso, villagers in Tuban on Java erected a special bundle of plants, called a *tumpeng robiyok* to bring prosperity for the new year. This bundle included traditional *tuwuhan* plants, such as banana, coconut, sugarcane, rice and areca nut. But to this group of plants are added other cultivated plants of a more recent origin such as maize, and cassava and various household utensils, baskets, strainers, ladles and even a water jar or *kendi*. The bundle can be seen as a reflection of traditional Javanese agriculture in the area.

Plants Used as Medicine

The use of wild plants as natural medicines has long been appreciated by many rural dwellers. Some 10 per cent of Indonesia's flora is regarded as having medicinal value and it is likely that future research and ethnobotanical surveys will reveal an even greater wealth of medicinal plants. Many of these are endemic to Indonesia and there is an urgent need to make objective and systematic assessments of their uses.

Dayak priests in Borneo have for centuries used plants as a way of exorcising sickness.

A poster promoting the various benefits of jamu, a traditional Indonesian herbal medicine.

»»Jamu sachets with elaborate packaging revealing the intensive marketing strategies for the various target groups.

From left: A woman in East Kalimantan collects leaves for curing stomach aches. A herbalist displays local jungle products. A jamu seller prepares liquid herbal medicine to be sold at about US 20 cents per glass. Colour plate of gutta percha (Isonandra palaquium), from Rhinos Vegetable Kingdom.

Traditional Healers

Traditional healers, or *dukun*, who use plants, herbs and spiritual powers to treat illnesses, are found throughout Indonesia. Although modern health centres, or *puskesmas*, are increasingly being used, especially by the younger generation, *dukun* still play a central role in village life. Their knowledge of medicines and healing practices belongs to an oral tradition which has been passed down from one generation to the next by word of mouth. This knowledge is not the sole preserve of the *dukun*, as many older people in Indonesia have at least a rudimentary knowledge of plant cures. The *dukun*, however, claim that a proper understanding of the treatment can only be acquired through divine inspiration. In this respect they occupy a privileged position in society, since it is believed that they are able to contact and learn from the spirits via dreams. Many *dukun* have an encyclopedic knowledge of plants and the diseases they can be used to treat. Botanists working with the Talang Mamak people in Sumatra, have found that traditional healers put their own taxonomic knowledge of forest plants to shame. *Dukun* may be male or female, although some female healers are called *bidans* or midwives. Recent investigations show that there is about one *dukun* to every 160 people in North Sumatra. This vocation often runs in families, with the sons or daughters of *dukun* learning the profession by assisting their parents. This is not always the case though and the role can be based on an assessment of an individual's 'spiritual powers'. In many cases the *dukun* has one or more apprentices. *Dukun* often possess unusual physical attributes — to be short-sighted or blind in this world, for example, often implies second sight in the spirit realm — and they are generally respected members of the community.

Medicinal Uses of Plants

Dukun tend to treat the signs and symptoms of an illness without using the sophisticated analytical techniques that western medicine uses to find the underlying cause. For the *dukun*, the physical and spiritual condition of a patient are often inextricably linked, and the relationship between causes, symptoms and cures is very different from that understood by modern medicine.

Traditional remedies have been arrived at largely through trial and error, with generations of human patients acting as experimental subjects. It is therefore significant that the medicinal uses of many plant species listed in ethnobotanical surveys from different areas are generally consistent, despite having been discovered independently by different and often widely separated cultures. Indonesian *dukun*, for example, use the same plant to treat malaria as do tribal people in the Philippines. This leads to the conclusion that some traditional cures must contain active and efficient ingredients.

All parts of the medicinal plant are used, either individually or in combination with other plants. The Kenyah Dayak people of central Kalimantan recognise and use roughly 200 medicinal and toxic species, representing 165 genera and 77 families. Not all medicinal plants are native to the region: some, of American or European origin were introduced long ago, either by design or accident.

Traditional healers obtain the ingredients of their cures from many parts of the forest, from among the fields created by slash-and-burn agriculture, and from semi-cultivated household gardens adjacent to their homes. Once collected, the *dukun* prepares the remedy, generally measuring out the plant in rough amounts such as handfuls, finger lengths or cupfuls. The plants are then ground, boiled, burned or mixed with food. The preparations are mostly decoction which are either drunk, or applied to a particular part of the body, commonly mixed with coconut oil.

In many parts of Southeast Asia one finds that medicinal plants are used in complex recipes containing many different species, with any one species being used to treat a variety of complaints. In Indonesia, however, most of the groups studied to date, with the exception of the Mentawai islanders, use a particular species for each complaint. Furthermore, the majority of recipes prescribed are simple: a single plant, or a combination of two or three plants, is the norm. The *dukun* maintain that this method makes it easier to detect the curative effects of a specific plant.

Conserving the Medicine Chest

Very little is known of the use of medicinal plants and traditional medicines in eastern Indonesia, as there has been no research into this since the early part of the century. It is, however, obvious that medicinal plant use is an important part of the traditional way of life in the region. This lifestyle has gone on for thousands of years and has ensured that people have lived in harmony with their forest homes. As the forest is threatened by development, so the knowledge of the *dukun* is being lost.

Many scientists believe that this traditional knowledge can speed the search for active chemical ingredients in plants. This could be the start of the development of modern medicines. But, many fear that this knowledge may be unfairly exploited and not result in the protection of the *dukuns'* home and their traditional way of life. A number of scientists around the world are working to ensure that at least a percentage of any profits which result from chemicals discovered in tropical forest plants will be used to conserve this unique and irreplaceable environment.

Jamu

A common sight in the streets of Javanese towns and cities is that of women selling *jamu*, traditional herbal medicines, from bottles in a basket strapped to their backs. When stopped by a customer, they find out what the person's ailment is — a headache, a backache, or simply a pick-you-up — and then, usually, cracking an egg into a glass, they add one of the liquids for the patient.

Various *jamu* are prepared by women from homegrown vegetables, supplementing them with ingredients purchased from a shop. It is estimated that up to 150 plants may be used in various *jamu*, including acacia eucalyptus oil, tamarind, and ginger as well as spices and sweeteners to add flavour.

It is generally claimed that *jamu* was first drunk by the palace ladies of the 17th century Mataram court to keep themselves beautiful. The recipes were (and still are) kept secret, being handed down by word of mouth. Today, some 80 per cent of the population of Indonesia will have taken *jamu* at some time. The commercial production of *jamu* began during the 1930s. Following World War II, the a broad range of *jamu* medications are produced.

Clockwise from top left: Dayak in the forest of West Kalimantan collecting cauliflorous fruits with medicinal properties.

Tacca integrifolia, a spectacular herb of the understorey vegetation of the Sumatran lowland rain forests, the root of which is pounded and eaten to treat amoebic dysentery.

The extraction of poison for blowpipe darts from its source in East Kalimantan.

PLANTS AS POISONS

Europeans first became acquainted with poisoned arrows and darts as soon as they reached Indonesia, but it was a long time before the poisons were traced to the plants which yielded them. Today, we know the chemical nature and actions of the cardiac glycosides that comprise the poisonous compounds. The principal poison used in the region is the latex of the upas tree (*Antiaris toxicaria*), shown below; a huge and widespread tree related to the fig. Another common poison is strychnine, which is obtained from large lianes of the worldwide genus Strychnos. Arrow poisons are usually concocted from the toxic compounds of several plant species, including from members of the *Tabernaemontana*, *Gnetum* and *Derris* genera.

The practice of catching fish using plant poisons, collectively known by the generic term tuba, is widespread in Indonesia. The sap, bark, wood, root and seeds of the leguminous climber known as akar tuba (*Derris elliptica*), are introduced directly into the water, which either kills the fish in the immediate vicinity, or else stuns them, thereby making them easy to catch. This indiscriminate method of fishing is now forbidden by most local authorities as its overuse has had a serious impact on the fish resources.

Resins

Resins are clear, volatile, organic liquids produced by trees and other plants, usually as a result of injury caused by wind, fire, lightning or human activity. With exposure to air these liquids lose some of their volatile components through evaporation, leaving a viscous residue which gradually hardens. Resins have been sought since antiquity for their use in the preparation of medicines, lacquers and varnishes; they continue to have a number of industrial applications in the modern world.

Colour plate of Styrax benzoin, *the main source of benzoin from the* Rhinos Vegetable Kingdom.

Minyak kayu putih or cajuput oil is a popular liniment distilled from the oil glands of the paperbark or whitewood tree, Melaleuca leucadendron.

»»A Dayak cleaning damar collected by the Ella Ulla river, West Kalimantan.

Camphor

There are many kinds of resin-producing trees in Indonesian forests but only a few are economically significant. Borneo camphor is one well-known product of the dipterocarp genus *Dryobalanops*. It comes mainly from *D. aromatica*, known locally as *kapur*. Camphor is formed from an oily secretion and occurs naturally as a crystalline substance in the wood; the amount produced varies between trees. Oil of camphor is obtained by cutting and firing a hole in the trunk.

The trade in Borneo camphor has an ancient history. Camphor was introduced to Europe via Arabia and historical records indicate that this trade dates back to at least the 6th century. There is also evidence of trade with China from the 8th century onwards. Interestingly, the collection of camphor was imbued with a mystical significance; the detection and collection of this product was often described using a special, arcane language developed from a corruption of local tongues. Nowadays Borneo camphor is no longer in demand, but *kapur* trees are still valued for their timber.

Dammar

The giant dammar tree (*Shorea javanica*) is another important resin-producing tree of the lowland rain forest. This huge dipterocarp grows naturally in the wild, but villagers in southern Sumatra have increased its abundance through planting and the removal of other tree species, so that dense stands have been formed. The resin forms in specialised canals which occur in the young wood. Stingless bees collect fresh dammar and use it to form their nest entrance tube. As it first runs from the tree, dammar is a stiff fluid. The outer surface hardens by the loss of volatile components, forming a crust often sculpted into the form of drops and ripples. Triangular holes, the size of a person's fist, are made in the trunk and the resin is scraped out every month. Tapping starts when the tree is about 20 years of age and can continue for another 20 years. A productive damar tree can yield up to 25 kilogrammes per month.

Dammar resin from *S. Javanica* and other related species of *Shorea* is used to make high-quality varnishes, cosmetics, pharmaceutical products, and paints. The highest grades are clear yellow in colour. The total damar production in Indonesia in the late 1980s was some 50,000 tonnes a year. While much of the resin is derived from *S. javanica*, commercially important resins are also collected, although less systematically, from other wild dipterocarps and also species of Burseraceae. Historically, the flame from burning dammar provided lighting in Borneo longhouses.

Minyak Keruing

Minyak keruing is an oily resin which traditionally was used for torches, caulking boats and medicines; it is still used as a base for certain perfumes. Like camphor and *dammar*, *minyak keruing* is obtained from species belonging to the Dipterocarpaceae family, in this instance trees belonging to the genus *Dipterocarpus*. The oil is extracted by hacking a large hole in the trunk and setting a fire in the cavity.

Sumatran Pine

Today, the most commercially important resin produced in Indonesia comes from the Sumatran pine (*Pinus merkusii*). When processed, the resin yields solid rosin and liquid turpentine. Rosin is used in paper-making and in the production of rubber, paint, varnish and printing ink; turpentine is widely used in the pharmaceutical industry and the manufacture of perfumes.

Originally a native of Sumatra, *Pinus merkusii* is now also planted extensively in Java and Sulawesi. Tapping starts when the tree is around 11 years old; a single tree may produce between 1 to 1.5 kilogramme of resin per annum. One hectare of trees can yield as much as 700 kilogrammes annually, and a single worker can tend two hectares, working part time. During the 1980s the Indonesian production of pine resin more than tripled. An extensive planting programme carried out during the same period has meant that an annual production of over 90,000 tonnes is possible. At present, however, there are a number of plantations in Sumatra and Sulawesi which are scarcely being tapped because the Sumatran pine is primarily grown for timber or woodpulp. In the future, however, resin production could be greatly expanded.

Manila Copal

Manila Copal was formerly of great commercial importance in the manufacture of linoleum, but is now used mainly for specialised varnishes. Known in parts of Indonesia as *damar minyak* it is produced by the bark of the giant coniferous kauri tree (*Agathis* spp.). Manila copal is obtained by cutting a V-shaped groove into the bark and collecting the exudate before it sets. Forests with abundant kauri once grew in southern Kalimantan and central Sulawesi and resin tapping was an important source of income for the indigenous people. At the peak of demand in the 1930s, many trees were killed by excessive tapping. Kauri is also extremely valuable as timber, and the forests where they once grew in such great profusion have now, for the most part, been felled. Lumps of 'fossil' Manila copal can be excavated from the ground below kauri forests.

Incense, Drugs and Dragon's Blood

Benzoin or gum benjamin, is a balsamic resin obtained from the wood of *Styrax* trees, mainly *S. benzoin*. It is fragrant and has various medical applications, as stimulating expectorant and for the treatment of skin irritations. Benzoin is also used as a fixative in perfumes. In order to collect the resin, the tree is wounded. This results in the development of new tissues containing resin-filled cavities which are then removed.

Other commercially valuable resins, produced by members of the Burseraceae family, occur as fragrant and non-fragrant varieties. The former is used in incense; the latter as an external medicine. A resin obtained from the fruit wall of a few species of the rattan genus *Daemonorops*, is extracted by shaking the fruits together in a closed basket. The residue is sold for medicinal use or as a dye in the form of a coarse powder known as dragon's blood or *jernang*.

The resin-soaked, diseased heartwood produced by certain species of *gaharu* — eaglewood or aloeswood (mainly *Aquilaria malaccensis*) — is burnt as incense and is still very much in demand in Indonesia and beyond. Demand for *gaharu* in the past has led to its virtual extinction on Siberut Island after a decade of intense exploitation. Elsewhere, *gaharu* trees occur only sporadically, scattered through the forest; they remain highly sought after as *gaharu* continues to command a high price.

Natural resins once played an important role in the manufacture of industrial and pharmaceutical products but have now been superseded by petroleum-based compounds. Recent reports indicate that drugs synthesized from the bark resin of some varieties of *bintangor* (*Calophyllum* spp.) have shown great promise in halting the replication of the AIDS virus under laboratory conditions. The genus *Calophyllum* comprises about 100 species and is widely distributed throughout Indonesia. Too small to be exploited yet for its timber, *bintangor* may yet prove to be of immense value in medicinal applications.

«« *A stand of large evergreen camphor trees (*Cinnamomum camphora*) from which camphor is collected.*

*Resin oozing from the stump of a freshly but probably illegally felled palanar (*Dipterocarpus littoralis*) in Nusa Kambangan, Central Java.*

Extracted resin, a secretion resulting from a fungal disease of the gaharu, also known as eaglewood or aloeswood.

✥ *A detailed view of the process of tapping gum from* Agathis.

Agathis *species in a plantation in West Java where trees are grown commercially for their resin.*

Rattan

Rattans are climbing palms which provide the raw materials for the cane furniture industry. While some climbing palms do occur in the New World, the true rattans are restricted to tropical and subtropical regions of the Old World. Over 600 different species, belonging to 13 genera, have been identified.

A close encounter of the nasty looking spines of the rattan, Plectocomia sp. *in the Kerangas (heath forest) of Pulau Bangka.*

»» *Workers unloading commercially harvested rattan downstream from the Bahau River in Kalimantan.*

A sharp contrast to the large scale commercial extraction of rattan is demonstrated by Dayak collectors of the West Kalimantan rain forest.

The World Market
Rattan species are particularly abundant in Indonesia, which dominates the world market in rattan sales, accounting for about 80 to 85 per cent of the world's supply. Their major habitat is tropical rain forest and in much of Southeast Asia they represent the most important forest product after timber. The trade in rattans and canes is worth approximately US$5 billion annually. The gathering of rattan is highly labour intensive, and employs some of the country's poorest people.

The Structure of Rattans
Rattans have long and very flexible stems that gain mechanical support from other trees and vegetation. In favourable conditions some species grow to great lengths, the longest cane ever recorded being over 175 metres long. Species are either single stemmed or multi-stemmed. The former can only be harvested once, whereas the latter can be harvested sustainably.

The stems of rattan palms are encased by leaf sheaths which are nearly always covered in fierce spines. Rattans also possess long whiplike growths, which are located either on the sheath or at the ends of the leaves. These whips, though armed with clusters of grapnel-like spines, do not serve as protectors of the plant. Instead, they play a major role in supporting the rattan as it climbs into the forest canopy. It is these terrible whips and spines that make the scientific collection of rattans so unpleasant and which are partly responsible for making this a poorly studied and, as yet, only partially, understood group of plants.

Harvesting from the Wild
Rattan is harvested by pulling the canes down from the surrounding forest and removing the spiny sheaths, leaves and whips, to expose the bare cane. This is a rather dangerous business as dead branches can be dislodged in the process, disturbing ants and wasps. The stripped canes are carried out of the forest and partially processed to remove excess moisture and to protect them from attack by wood-boring beetles. Small diameter canes are dried in the sun and often smoked over burning sulphur; large canes are boiled in oil — usually a mixture of diesel and palm oil. They are then sold to middlemen who convey them to the furniture manufacturers.

Maintaining Supplies
In the past, much of the commercially harvested cane was exported to manufacturers in Singapore, Hong Kong, Japan, Europe and North America. Indonesia has now introduced export bans on raw cane to encourage the manufacture of rattan furniture within the country. This adds to the value of the exported product, while also helping to conserve existing supplies of wild rattan. Indonesia's restriction on the export of raw cane has, however, put extreme pressure on the natural cane resources of other countries where there are no controls. The result has been severe over-exploitation and, in some places, the disappearance of natural cane resources altogether. At the same time, the worldwide destruction of rain forests has greatly diminished the natural habitat of rattans over the past few decades and there is now a very real shortage of supply.

Conservation of Rattans
In the mid-1970s, forestry departments in Southeast Asian countries became aware of the vulnerability of rattan and began investigations aimed at safeguarding the long-term supply of canes for the industry. Cultivation seemed to present the best hope for the future and early research examined existing methods. In one small area of Kalimantan, rattan has been cultivated in permanent rattan gardens adjacent to rivers that are subject to severe and prolonged flooding. This land is more-or-less unsuitable for any other form of agriculture because of these periodic inundations and the very high acid content of the soil, yet appears to be ideal for the cultivation of the species, *Calamus trachycoleus*.

Elsewhere, in another part of Kalimantan, *C. caesius*, has been cultivated during the fallow period of the shifting cultivation cycle. Rattan seedlings are planted after the rice has been harvested and the canes allowed to grow in the secondary forest that colonises the previously cleared land. After a period of 14 to 15 years, the farm owner returns, harvests the rattan, fells the secondary forest, burns the field and begins the cycle again with another rice crop. These local methods of cultivation have subsequently been used as a model elsewhere.

Calamus trachycoleus and *C. caesius* are small diameter canes (six to 12 millimetres) of excellent

PEOPLE AND PLANTS: FOREST PRODUCE

quality, but it is the large diameter canes (in excess of 18 millimetres) which are needed for the framework of rattan furniture. The main focus of research has been to find large diameter canes that are suitable for the domestication and cultivation in a variety of habitats. With over 600 species to choose from, and a huge geographical, altitudinal and ecological range, determining the right cane for the right habitat is clearly a complex process. More detailed information on the ecology and classification of rattan is required to further develop this resource and reduce pressure on threatened species. Particular attention is now being given to selected large diameter species such as *C. manan*, *C. zollingeri* and *C. subinermis,* which show particular promise for future cultivation.

Commercial Planting

In recent years, new commercial rattan plantations have been established in Indonesia and elsewhere in Southeast Asia. Growth rates in the new estates have been most encouraging: the best large diameter cane, *C. manan*, and the small diameter *C. trachycoleus* have both been recorded as growing at more than six metres a year. Several plantations outside Indonesia have already reached harvestable age and the financial return from these estates seems encouraging. There are also some unexpected benefits to be had from these plantations. In order to grow properly, rattan has to be planted under some sort of tree cover, such as logged–over forest, secondary forest, fruit orchards, tree plantations or even rubber estates. The planting preserves tree cover and, in places where this is provided by semi–natural forest, the local wildlife is also preserved. One of the highest population of orang utan in Borneo is now found in a rattan estate.

Commercial planting offers attractive prospects for wise land use, allowing a crop to grow with minimal disturbance to the existing vegetation. The potential as a smallholder crop is even greater. Small diameter species can be planted on a small scale under fruit trees or in rubber gardens, allowing the independent smallholder to gain extra cash returns while also reducing pressure on surrounding forested land.

THE VERSATILE RATTAN

❶ *Calamus manan*, the most sought after large–diameter cane, is quite widespread, usually confined to the steep slopes of hill dipterocarp forest of Sumatra and southern Borneo.

A The lateral inflorescence

B Section of the woody stem heavily covered with spines.

❷ to ❹ The highly versatile rattan manifests itself in a wide range of products including furniture, as in the chair, and basketry as in the two baskets used as backpacks for transporting harvested pepper to the market.

❺ It can even be used as clothing, as worn here by a Yali man, in the form of rattan loops.

Bamboo

Bamboo is of great importance to the people of Southeast Asia, where a number of naturally occurring species are exploited for a wide range of purposes. In Indonesia, bamboo is intimately associated with rural life: in Java, for example, one typically finds clumps of bamboo planted around villages so as to be readily available for everyday use.

Attractive indigenous and naturalised bamboo culms (from left): Schizostachyum brachycladum, Phyllostachys nigra *and* P. aurea.

Bamboo manifests itself as (top to bottom): a bridge, scaffolding on a modern building and as the pole and stool of a street vendor.

A traditional Toraja house in Sulawesi with an elaborate bamboo roof structure using bambu lemang. Details of the weave are seen in the inset.

Characteristics of Bamboo

Bamboos are members of the grass family, Gramineae, and form the tribe Bambusae of the sub-family Bambusoideae. They share some common features with other grasses, including the structure of their flowers: these are very small, just two to seven millimetres long, with two or more florets borne along a jointed branch forming a spikelet. These differ, however, from flowers of other grasses in that they possess three lodicules (tiny structures thought to be equivalent to sepals and petals) and six (or occasionally three) stamens in each flower. Bamboos are more obviously distinguished from other grasses by their woody, branched stems or culms.

Bamboos, like other grasses, are rhizomes. These form complex, branching, subterranean systems which are the foundation of the plant. The bamboos found in Indonesia have a sympodial rhizome system, in which the basal part of each rhizome is short and thick with the apex growing upwards to become a new culm. A bamboo culm is divided into nodes and internodes, with the diameter ranging from one to 20 centimetres. These culms grow to between seven and 30 metres in height and, depending on the species, may either be erect with drooping tips, or derive mechanical support from their surroundings as evident by scrambling or climbing varieties.

Distribution of Bamboo

Bamboos are found all over the world, but the majority occur in the tropics, where they grow wild at low to medium altitudes. Globally, there are some 70 genera of bamboos, comprising some 800 to 1,000 species, of which 60 to 100 species can be found in Indonesia. Four of the genera, *Bambusa, Dendrocalamus, Gigantochloa* and *Schizostachyum*, include species that are now widely cultivated. Representatives of the other four indigenous genera have a more restricted distribution, and are generally found in montane or mossy forest habitats.

The Many Uses of Bamboo

Bamboo culms have many characteristics that make them suitable for a variety of uses. Diverse species are regularly employed for building materials, in basketry and as a vegetable. Some species are only used as building materials: as pillars and posts, or in the construction of walls, roofs and floors. When used for pillars or other load-bearing structures, the culms should have a large diameter with thick walls and relatively short internodes. Suitable species include *bambu duri* (*Bambusa blumeana*), *bambu betung* (*Dendrocalamus asper*), *bambu gombong* or *andong* (*Gigantochloa pseudoarundinacea*) and *G. robusta*. Medium sized culms that have relatively thin walls are suitable for floors, roofs and walls.

Bamboo walls are made by cutting the culms to an appropriate length, splitting them along one side, and then flattening them out. These lengths can either be used directly as vertical strips, or else woven into larger sections. Alternatively, the culms may be split into very thin, long strips which are then plaited into large panels. These are used for partitions and ceilings as well as walls. In roof construction, the bamboo culms are split into two and arranged in such a way that they resemble corrugated iron.

Indonesians generally use baskets and kitchen utensils made from species with smaller diameter culms and relatively thick walls, yet which split easily. *Bambu tali*, or *apus* (*Gigantochloa apus*), is planted extensively in villages in Java for this purpose, but may also be used in construction as pillars.

Culms are also used as fruit and vegetable props, carrying poles, fishing rods, outriggers, boating poles, posts and fences. Furthermore, because they do not produce an unpleasant odour, culms are used as containers for water, pipes, conduits and troughs; for storing palm wine and other beverages, and as a receptacle for cooking rice, vegetables and meat. Fish traps are made of split bamboo joined together with strips of rattan or bamboo.

Bamboos are also used for making various musical instruments. Flutes are made from the

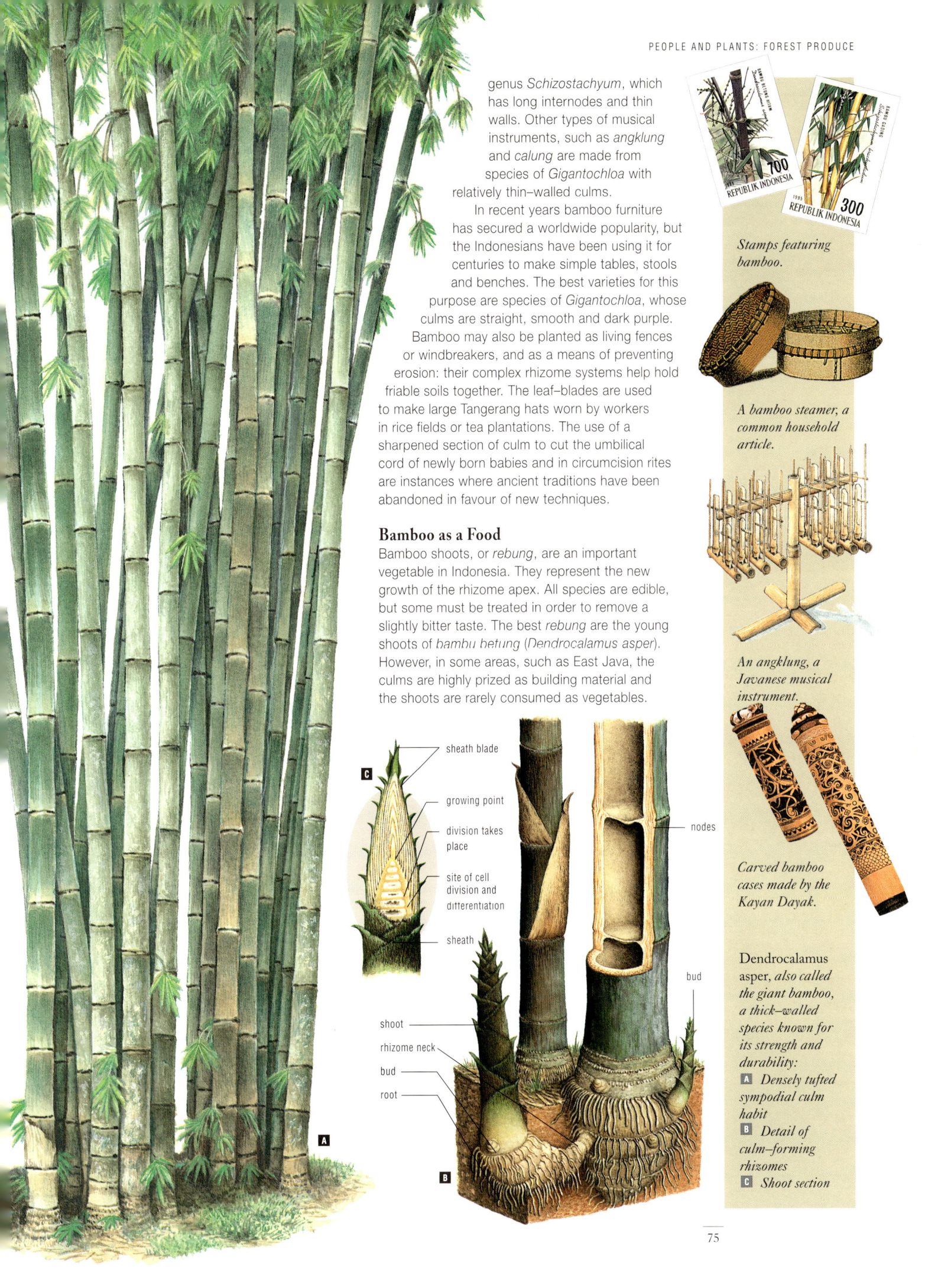

PEOPLE AND PLANTS: FOREST PRODUCE

genus *Schizostachyum*, which has long internodes and thin walls. Other types of musical instruments, such as *angklung* and *calung* are made from species of *Gigantochloa* with relatively thin–walled culms.

In recent years bamboo furniture has secured a worldwide popularity, but the Indonesians have been using it for centuries to make simple tables, stools and benches. The best varieties for this purpose are species of *Gigantochloa*, whose culms are straight, smooth and dark purple. Bamboo may also be planted as living fences or windbreakers, and as a means of preventing erosion: their complex rhizome systems help hold friable soils together. The leaf–blades are used to make large Tangerang hats worn by workers in rice fields or tea plantations. The use of a sharpened section of culm to cut the umbilical cord of newly born babies and in circumcision rites are instances where ancient traditions have been abandoned in favour of new techniques.

Bamboo as a Food

Bamboo shoots, or *rebung*, are an important vegetable in Indonesia. They represent the new growth of the rhizome apex. All species are edible, but some must be treated in order to remove a slightly bitter taste. The best *rebung* are the young shoots of *bambu betung* (*Dendrocalamus asper*). However, in some areas, such as East Java, the culms are highly prized as building material and the shoots are rarely consumed as vegetables.

Stamps featuring bamboo.

A bamboo steamer, a common household article.

An angklung, a Javanese musical instrument.

Carved bamboo cases made by the Kayan Dayak.

Dendrocalamus asper, *also called the giant bamboo, a thick–walled species known for its strength and durability:*
A *Densely tufted sympodial culm habit*
B *Detail of culm–forming rhizomes*
C *Shoot section*

Timber: Species and Kinds

Timber is one of Indonesia's major export commodities along with oil and gas. Over 95 per cent of the wealth generated by forests today comes from their timber resources, a notable contrast with earlier times when resins, rattans and other forest products were considerably more important. Nearly all the timber produced in Indonesia still comes from logging virgin forests.

COMPOSITION OF A TRUNK
- Pith
- Heartwood
- Sapwood
- Cambium
- Inner bark
- Outer bark

Koompassia excelsa, a very tall tree grows up to 85 metres. With its distinct buttressed columnar trunk, it provides a medium–heavy timber used as structure support and firewood.

Paper pulping is an increasingly important component of the Indonesian timber industry.

Western Indonesia's Forest Wealth

The forests of western Indonesia are amongst the richest in the world in terms of their diversity of plant and animal species. They are also amongst the most valuable in as far as their timber resources are concerned. The most important timber regions in Indonesia are in Sumatra and Kalimantan, where the forest is dominated by large trees of the Dipterocarpaceae family. Dipterocarp species are among the most valuable to the timber industry, and their abundance allows over 100 cubic metres of timber to be extracted per hectare in parts of Kalimantan. Elsewhere, the yield may be as much as 80 cubic metres, but 20 to 30 cubic metres per hectare is the average figure for eastern Indonesia and most other tropical areas.

Indonesia's timber industry is dominated by tropical hardwoods, i.e. timber which is produced by flowering plants, or angiosperms. In Indonesia softwoods, the timber produced by conifers, is relatively unimportant both in terms of the volume extracted and its commercial value. The principal softwoods from Indonesia are *Araucaria* from Irian Jaya and the *kauri* (*Agathis*). Both of these are extremely valuable and much sought after, for its timber and *Agathis* also yields the resin *damar minyak* or Manila copal. Once common in large areas of Kalimantan and Sulawesi, *kauri* has been heavily exploited and now only occurs in small pockets.

In timber commodity terms the hardwoods are divided into two kinds, namely Heavy Hardwoods, which are dark, dense, and naturally durable, and Medium plus Light Hardwoods, which are pale in colour, of medium to low density and are mostly not naturally durable. The former include cabinet woods which have an exceptionally pleasing appearance. Some Heavy Hardwoods, such as *bakau* (*Rhizophora*), which is obtained from mangrove forests, are used for the construction of bridges, buildings and wharves; others, such as *merbau* (*Intsia* spp.), are used for flooring. Heavy Hardwoods may be difficult to work if they contain silica. Conversely,

EIGHT TIMBER TYPES SHOWING DIFFERENT GRAINS

Top row: (left) *Xylia xylocarpus*, a cabinet wood, native of Africa and rare in plantations, (right) valuable burr–figured piece *Pterocarpus indicus*, the only native species used as a cabinet wood; Second row: (left) normal *P. indicus*, (right) *Albizia*, a light hardwood; Third row: (left) *Sindora*, a light hardwood, (right) *Koompasia*, a heavy hardwood; Bottom row: (left) *Intsia palembanica*, a heavy hardwood, (right) *Dalbergia* or rosewood, a cabinet wood grown in plantations.

Light Hardwoods are easy to work and are therefore used for interior construction and furniture or else are converted into plywood.

Indonesia banned log exports in the mid–1980s and there is now little demand for Heavy Hardwoods which are therefore seldom cut today. By contrast, the Light Hardwoods market remains buoyant and Indonesia is one of the world's main plywood producers as well as one of the world's largest source of sawn timber. Most plywood is made from *meranti* (*Shorea*) and *mersawa* (*Anisoptera*), which both belong to the dipterocarp family. Mengkulang

STATISTIC OF PAPER AND PAPERBOARD (1985 – 1994)

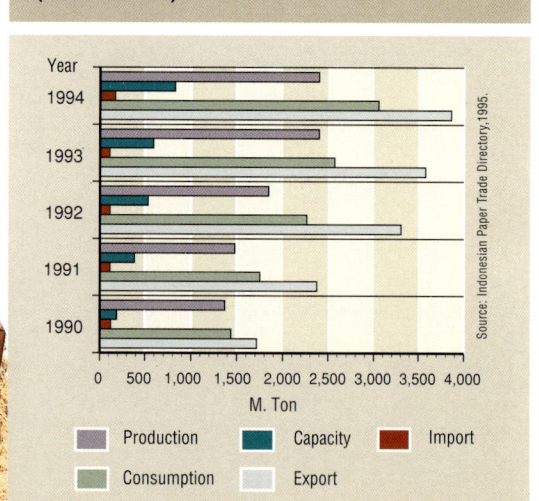

Source: Indonesian Paper Trade Directory, 1995.

Legend: Production, Capacity, Import, Consumption, Export

*Ironwood (*Eusideroxylon zwageri*) forest at Rimbo Kilm in the province of Jambi, central Sumatra. Left, detail of ironwood flowers.*

The small pinkish white teak flowers are pollinated by insects. The flower has a one–day life cycle.

(*Scaphium*), is also frequently used, as is the softwood *kauri* (*Agathis*) when available.

Paper Pulp

In addition to its use in the construction and plywood industries, Indonesian timber is also increasingly being employed in the manufacture of paper pulp. The principal source of Indonesian pulpwood comes from plantations of *Acacia mangium*, which originally was a native of Australia, but is now grown commercially in Indonesia. Another source is the native Sumatran Pine (*Pinus merkusii*), which is planted around Lake Toba in north Sumatra. Large areas of residual natural forest, after timber extraction, are also being cleared for pulpwood and the land is then planted with trees grown for plywood.

Teak

Java is one of the world's major suppliers of teak. Teak or *jati* (*Tectona grandis*) is a highly durable, reasonably hard, yet easily worked timber. The extensive teak forests that are found in the seasonally dry parts of Java almost certainly originated from seeds brought from India by early Indian sailors and merchants. Teak trees can grow up to 50 metres tall, with a diameter of 125 centimetres or more, but is usually smaller than this. Teak flourishes in climates where there are four to seven dry months in the year and can be found at altitudes which range from sea level up to 700 metres in elevation. The tree produces large panicles of whitish flowers which open a shortly after sunrise, and have an optimum pollination period around midday, each flower only lasting one day. Pollination is effected by insects, the most important being bees. Seeds develop to their full size some 50 days or so after pollination, but are ripe only after 120–200 days. The natural fruit dispersal agent is believed to be strong, pre–monsoon winds. The ground litter of large leaves and branches decays slowly and suppresses the growth of other plants. It also forms an inflammable tinder, but when occasional fires sweep through the forest the teak trees remain undamaged. Other tree species, however, succumb and the soil surface may be subject to erosion.

Future Prospects

As the days of the old forest giants draw to a close with the impending exhaustion of the virgin rain forests and their large stocks of well–known timbers, the market will begin to accept smaller trees and a wider range of species. Plantation–grown timber is also gaining importance. Old rubber trees (*Hevea brasiliensis*), with their pale wood and attractive grain, are already being used for furniture. Large laminated planks are built up from small blocks, using modern glues. This technology opens up the possibility to utilise numerous small native trees. There is an increasing demand for pale, low density timbers such as *sungkai* (*Peronema canescens*). Sungkai is small, fast-growing and has very attractive grain, resembling its larger relative, teak.

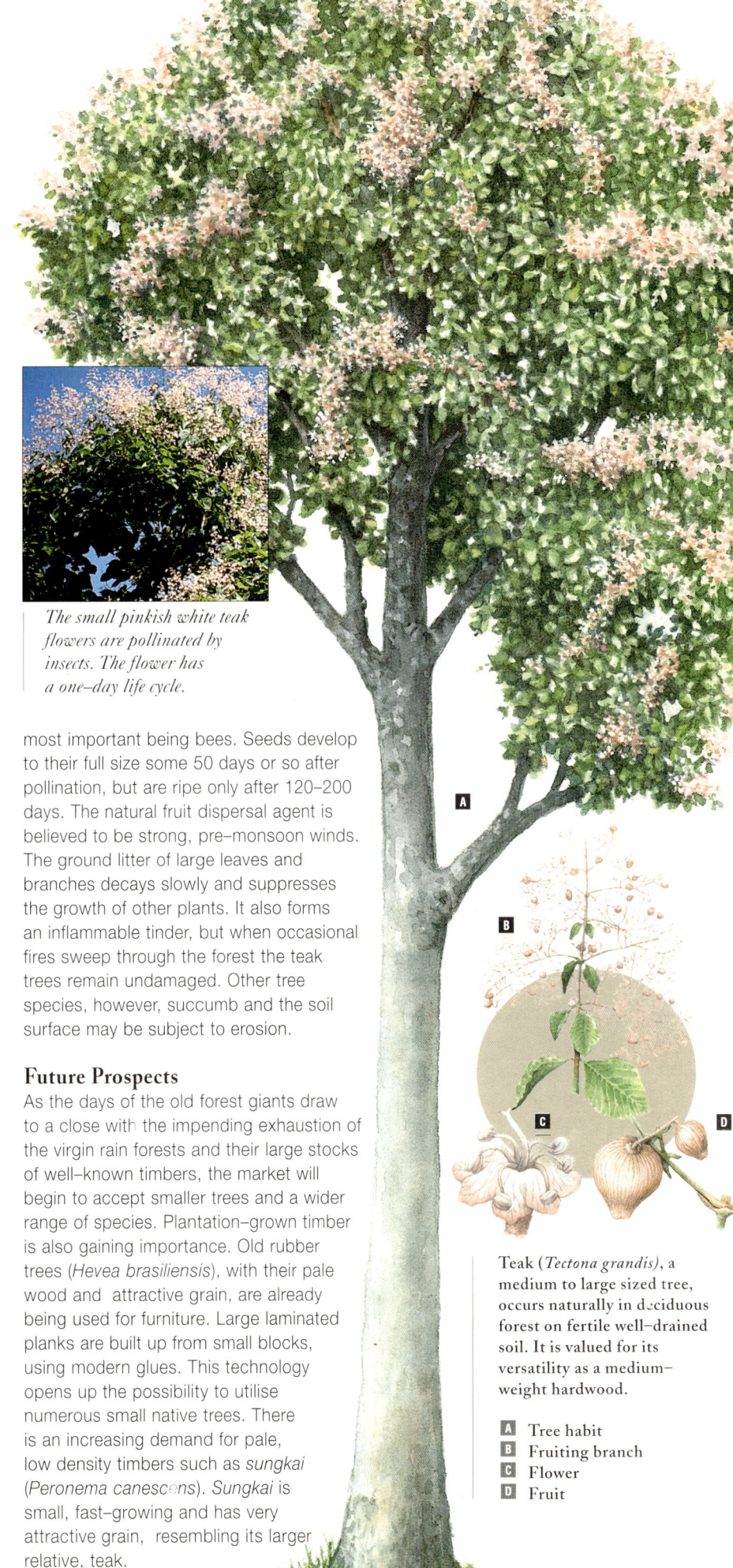

Teak (*Tectona grandis*), a medium to large sized tree, occurs naturally in deciduous forest on fertile well–drained soil. It is valued for its versatility as a medium–weight hardwood.

A Tree habit
B Fruiting branch
C Flower
D Fruit

Sustainable Use of Indonesian Timber

When the International Union for Conservation of Nature (IUCN) decided on a new definition of what it means to use wild species in a sustainable manner, an international consensus was sought. Hundreds of scientists and resource managers were consulted over several years, but what the definition they arrived at ultimately revealed was just how difficult it is for wild species to be used in a truly sustainable way.

»»Logging trucks with their heavy load of freshly cut timber, leaving behind a trail of dust, are a common sight in logged areas of northeast Kalimantan.

Sustainable Management

The key feature of sustainable use and management is that the use of wild species is deemed to be sustainable only if it does not reduce the potential of that species to be used in the future, or else impair the long-term viability of that species or any other, or that of the ecosystems which support and depend upon those species. This definition is made even more rigorous when the meaning of the term 'viability' is explained. When applied to a given species, it means that no harm is done to the capacity of that species to maintain its genetic diversity, its potential for evolution, and its likelihood of survival. When applied to an ecosystem as a whole, it means that no harm is done to the capacity of the ecosystem to maintain the diversity of its habitats and genetic lineages, or to its capacity for continuity and renewal, or to its productivity in terms of natural resources.

There is a world of difference between these cautious phrases and what actually happens in a rain forest when it is logged: the roar of bulldozers; the churned red mud; the puddles of diesel oil; the howl of chainsaws; and the crashing fall of great trees. There is no chance that such mayhem could ever be made to fulfil the IUCN's criteria for sustainable harvesting. It is just too uncontrolled and rampantly destructive.

The devastating results of the wholesale logging of primary forest.

Workers processing timber logged in the vicinity of the Samarinda Sawmill in East Borneo.

»»The modern method of felling of trees: East Kalimantan.

Weighing up Financial Costs and Rewards

Controlled timber extraction is extremely expensive, both in terms of the labour costs and the technology required. Or these are regulated to the point of being virtual government monopolies. The only logging systems which come close to being sustainable are either very labour-intensive, using pitsaws and headloads, or require expensive technology such as helicopters to pluck timber from remote locations. Such strategies are unlikely to find a place in the policies of a modern corporation; when forestry is managed as a business, everyone in a logging company has a strong interest in keeping production costs to a minimum, which means avoiding investing in controlling the logging process. Furthermore, in almost every instance, logging companies have no secure tenure, and therefore have no long-term incentive to protect the ecosystems that they exploit.

The demands of shareholders and creditors are paramount in this capital-intensive business, where money in hand is worth more than potential earnings in the form of well-managed forests for the future. A final concern is that in most instances the environmental costs of logging are externalised and are not included in the price of the timber harvested. That is to say, the economic value which is lost from a forest as it is logged, for example, and the benefits provided by its watershed role, or its ability to supply wild meat and fish to local people — are simply not taken into account and are therefore not reflected in the sale price of the timber. Consequently, neither the timber company nor those who buy its products pay for the damage to the environment, the costs being borne instead by the public, which is largely unaware of these. Given this, it is hard to imagine a combination of factors which are less likely to result in the sustainable management of tropical rain forests.

PEOPLE AND PLANTS: FOREST PRODUCE

After the logged timber is transported out of the forest, it is sent downstream in batches for processing. The logs are guided along the water route, as seen here in Kalimantan, by small boats.

Evidence of heavy erosion — deep gullies formed as a result of deforestation where the protective forest cover is removed, exposing the soil to the elements.

A view of a primary forest patch — fragmented by the construction of logging trails for the transport of felled timber.

Qualified Successes

The task of imposing higher standards on the timber industry, against its will and contrary to its financial interests, is usually the job of civil servants within a country's forestry department. These professionals are in a very frustrating position: anxious on the one hand to see logging performed to the standards to which they were trained, they are nevertheless caught up in a system which is dedicated to avoiding compliance with these standards. At the same time, often, unsustainable levels of logging play an integral role in the process of national development, as the timber industry creates wealth for investment, and consequently, government revenue through taxation and employment.

In these circumstances, it is hardly surprising that the advocates of sustainable harvesting have had only a limited impact on the developing world's timber industries. Some gains have been made, but have involved compromises as continued pressure from professional foresters has forced governments to reconsider their policies and led to the tightening up of laws and the capacity to enforce them. This has raised the price of poor logging techniques, making it worthwhile for companies to invest more in controlled timber extraction. Meanwhile, environmental groups which have lobbied long and hard, have had their own effect on policy–making, while better publicity for the issues involved has also influenced the complex dealings and operations of major timber companies.

This process adds up to a gradual reappraisal of the kind of logging which is considered acceptable by forest departments and logging companies. The new consensus still remains far short of achieving sustainability, but nevertheless represents a marked improvement. The eventual outcome may take the form of a set of standards, designed to achieve minimum ecological damage for a given level of investment in new logging strategies, and which incorporate the economic and environmental costs of logging into the price of timber. To be useful, however, these measures must be realistic in so far as they can be achieved by at least some timber producers.

The standards will be based on compromises around a set of simple rules known by foresters for decades. These may, one day, influence the behaviour of the timber industry as a whole:

- protect forests from unplanned use, theft, or conversion to farmland;
- assess the annual cut realistically;
- annual felling zones to be clearly demarcated;
- trees to be marked for felling or retention;
- detailed pre–felling inventories;
- checking the harvest to prevent only valuable timber from being selected;
- continuous inventory and monitoring of changes;
- maintaining main roads to control erosion on spur roads and skid trails.

Applying these principles demands discipline, commitment and a long–term perspective.

79

The Challenges of Indonesian Forestry

Indonesia is a major producer of tropical timber and wood products. It has a land area of about 1.9 million square kilometres, of which nearly 1.2 million remain forested. Commercial logging began in Kalimantan during the early 1960s, but by the late 1980s an estimated 435 million cubic metres of timber had been extracted from Indonesian forests. Such levels of harvesting cannot be maintained and new measures must be adopted to protect remaining forests from over-harvesting.

A great deal of construction timber is produced in small sawmills such as this one in North Sulawesi.

Timber Export

Commercial logging began in Kalimantan during the early 1960s, and within ten years had become widespread there as well as in Sumatra and Sulawesi. Today logging concessions cover more than one-third of Indonesia's forest land. Much of the timber extracted from these concessions is for export — about 10.5 million cubic metres of wood were exported from Indonesia in 1991. This is more-or-less the same volume per year as in the previous ten years, but the form in which the timber is exported has changed greatly in that period. In 1981, about eight per cent of all wood exports were in the form of plywood but, by 1991, that figure had reached nearly 80 per cent. Within a single decade, Indonesia had been trans-formed into the world's largest producer of plywood, using its huge supply of giant, straight-boled dipterocarps and other trees ideal for 'peeling' into sheets by spinning their trunks against immense mechanical blades.

It seems likely that the export market will continue to change over the next few decades, as the domestic market takes an ever-increasing share of the plywood industry's output.

A Threat to the Environment?

As the virgin forests of Kalimantan and Sumatra have steadily been logged over, the supply of raw materials has begun to decline. Indonesia's next great vision is to become the world's largest wood-pulp and paper producer. It is a vision which the efficient timber industry seems well able to turn into a reality. The determination and vigour with which Indonesia harnesses its natural resources to the engine of economic growth is truly phenomenal.

But how will this affect the environment? Although Indonesia occupies only about 1.3 per cent of the world's land area, it is estimated that as many as 17 per cent of all the plant and animal species on the planet occur within its national boundaries. Indeed, by any measure of species richness and endemism, whether it be plants, birds, reptiles, amphibians or mammals, Indonesia is consistently listed among the top few nations in the world. This great biological diversity is linked to the fact that Indonesia's 17,000 islands lie within several major biogeographic regions spanning between Asia and Australia, and the Pacific and Indonesian oceans. Each of these regions has different evolutionary connections and climates, which has encouraged and sustained such a high degree of species diversity. Even the seas between these islands and land masses are the most species-rich in the world.

Conservation Measures

The development of Indonesia's timber industry has been based on dividing forest land according to the intended function of each area. The main categories are: nature reserve and protected forest (where no timber extraction is allowed); limited production forest (for non-industrial selection felling); regular production forest (for industrial selection or for clear-felling and replanting, according to forest type); and conversion forest (for clear felling and

FOREST AREA
Forest 1,179,140 sq. km
Non-Forest 739,523 sq. km

Source: The Conservation Atlas of Tropical Forests – Asia and the Pacific, London, 1991.

PLYWOOD EXPORTS (in thousand m³) FROM INDONESIA

MARKET	1988	1992
Japan, China, Taiwan, Korea	72	300
Europe	99	270
America/Canada	33	136
Sinagapore/Middle East	59	241
Others	16	48

Source: Food and Agriculture Organization (FAO) of the United Nations, Rome, Italy, 1993.

FOREST LAND USE BY CONSENSUS (in Ha)

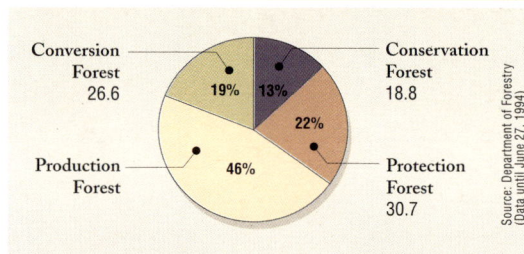

- Conversion Forest 26.6 (19%)
- Conservation Forest 18.8 (13%)
- Production Forest (46%)
- Protection Forest 30.7 (22%)

Source: Department of Forestry (Data until June 27, 1994)

The guidelines for determining which land should be state forest and the purpose to which it should be put were completed in 1982 following government regulation No. 33/197 (as a regulation for implementing Basic Forestry Law No. 5/1967), with the consensus of all relevant ministries. Five categories of the forest are recognised.
- **CONSERVATION FOREST:** designated for nature and genetic conservation, in which no exploitation is permitted.
- **PROTECTION FOREST:** designated for water and soil conservation, in which no exploitation is permitted.
- **LIMITED PRODUCTION FOREST:** designated for soil erosion prevention and timber production, in which selective felling is permitted.
- **PRODUCTION FOREST:** designated for selective felling or for clear felling and replanting.
- **CONVERSION FOREST:** designated for conversion to agriculture or other uses in which clear felling is permitted.

With the exception of CONSERVATION FOREST, these categories were defined taking into account each area's vulnerability to soil erosion, based on soil type, average slope and rainfall intensity.

conversion to other uses). This approach is, in theory, a good way to ensure that forest lands are used sensibly. In reality, however, many areas were allocated to one category or another long before sufficient information was available to justify these decisions. As a result, some regrettable errors have occurred resulting in serious environmental problems. For example, valuable forests were cleared in the wrong places, while in other instances, logging roads were built through vital nature reserves to give access to production forest elsewhere.

The pragmatic Indonesian government has long recognised these difficulties, and has continued to improve the system by commissioning forest policy studies, and investing in detailed mapping operations and assessments of land capability. The race to apply rational planning to the use of Indonesian forests has often been outflanked, however, by unregulated deforestation. Excessive logging, agricultural encroachment, transmigration programmes and forest fires have all played a part in this process. The task of protecting Indonesia's conservation areas has been hard because the conservation department (Perlindungan Hutan dan Pelestarian Alam) has had few resources at its disposal. Nevertheless, about 25 per cent of the country's land area has been set aside for the preservation of the nation's biological diversity and to maintain specific types of ecosystem intact.

This national system of conservation areas is adequate on paper to ensure the survival of most of Indonesia's biological resources. The next stage of development will tell whether the great skill and dedication of which Indonesians are capable, can be applied to the preservation of the country's immense wealth of natural species indefinitely.

A Critical Decade

Indonesian foresters have seen Sumatra, Sulawesi, Kalimantan and Irian Jaya as sparsely-populated frontiers, laden with unclaimed timber resources ripe for harvesting. Over-logging and forest clearance have taken a terrible toll of the country's natural ecosystems, but the scene is now set for a more controlled and sustainable use of forests. If Indonesia is to enter the next century with abundant natural forests and biodiversity resources, reforms are needed in this decade to build on the progress that has already been made by Indonesian forests.

The destructive effects of logging on the existing vegetation causes not only heavy wastage, but also compacts the soil by creating timber tracks.

Clockwise: Whole stands of tropical rainforests are cleared to make way for industrial timber plantation. This results in dramatic changes to the landscape.

A worker in East Kalimantan seen manoeuvring meranti logs with the help of heavy duty equipment such as a tractor.

Muddy rivers in Kalimantan are indications of serious erosion taking place further upstream.

(Left). An early form of cultivation is demonstrated by the planting of bulbous plants with dibble sticks by the people of the Island of New Guinea (now Irian Jaya).

(Right). Plantation workers tapping rubber and then transporting the tapped latex.

Bottom: Rice fields in the hill terraces of Bali showing one of the three main landscape types where rice farming is practised. In the foreground a farmer is seen balancing two baskets with a pole, used as a transportation tool in the rice terraces.

PEOPLE AND PLANTS: CULTIVATION

The earliest cultivation may have been entirely fortuitous, as when durian trees develop after their seeds are tossed into the undergrowth when a person has finished feasting on the delectable pulp. Wherever early people settled, whether in caves or rude shelters, they would probably have brought seeds with them, or transplanted young plants from the forest in order to avoid long journeys to collect a useful product. Today, a multitude of crops are grown in Indonesia, many of which have been improved by the deliberate selection of desirable traits.

Relatively few of the plants cultivated in Indonesia today are actually native to the region. Those that are include yams, mangoes, mangosteen, rambutan, cloves, some citrus fruits, wild cinnamon, nutmeg, petai, sugarcane, sago and coconut. Some cultivated species, such as the betel palm and sweet potatoes, have no known native range, having always been found only around human habitation. Common introduced species include rice, soya, other citrus fruits and tea, all of which probably came from East Asia; potatoes, tobacco, mahogany, rubber, maize, cassava and cocoa, which were brought from the Americas; coffee and oil palm from Africa; and pepper, common cinnamon and teak from South Asia. Numerous varieties of these species exist, each with a specific taste, size, productivity, or seasonality. It seems likely that varieties may have become extinct as better versions were found or developed by selection. Today, rural areas, remain an important reservoir of genetic variation of many more wild plants which might yet prove useful.

There is a wide range of cultivation patterns in Indonesia. In parts of Sumatra and Kalimantan the forest has been enriched by the selective removal, tending and planting of different species. Shifting or swidden cultivation is practised in other areas. In the Mentawai Islands, Maluku and the lowlands of Irian Jaya, sago palms provide the main staple, with the trees forming their own swamp by trapping water between their dense roots. Javanese culture has encouraged the development of home gardens in which useful plants are planted in and around houses.

Ornamental plants are also widely cultivated, both on a small–scale in village homes and in urban plots, producing flowers for export. Whether or not the species and varieties are native, the climate is conducive for strong and rapid growth, and Indonesian homes are often surrounded by a kaleidoscope of colour.

Origins of Cultivation

Indonesia does not appear to have been an ancient or original centre for the development of agriculture. There is, however, little doubt that many cultivated and domesticated plants which are economically important today, have a long history of cultivation in the region.

»»Scanning electron microphotograph of rice chaff from Ulu Leang I in South Sulawesi.

Prehistoric burned rice dating to about 4,000 BC from cave deposits in South Sulawesi.

Excavating a small trench at Ulu Leang I, in South Sulawesi.

Origins of Rice Cultivation

Asian rice is the most important of the world's food grains, both in terms of the number of people it feeds, and the intensity of its cultivation. Rice first became important as a cultivated crop in the middle and lower Yangtze Valley of central China some 10,000 to 7,000 years ago. Its earliest appearance in Southeast Asia dates back 5,000 years, but it did not become common in the region until some 3,000 to 4,000 years ago. It seems that the development of the main regional varieties took place as people extended the plant's natural climatic range northwards into temperate China and Japan, eastwards into semi-arid western India and Pakistan and southwards, through the equatorial belt, into Indonesia. It is now believed that Asian rice evolved from a wild annual ancestor, called *Oryza nivara*, which grew in the region extending eastwards from the central Ganges Valley to the South China Sea.

Rice (Oryza sativa), in full flower.

Rice in Indonesia

Over the past 25 years, there have been numerous finds of rice at archaeological sites in Asia. Efforts have been made to screen archaeological deposits for burnt rice and to recover rice grains from the clay matrix of pottery and bricks. The remains can be categorised according to whether they are wild or domesticated and, if the latter, by the variety. To date, much more effort has gone into this type of research than into trying to understand early rice cultivation systems and the effect of these on social and cultural change. Even so, remarkably little palaeo-botanical data has been recovered, precisely identified and dated.

Unfortunately rice pollen is indistinguishable from the pollen of most other grasses, so pollen fossils held in core samples of peat taken from a number of Indonesian lakes cannot help date the arrival of rice in an area. Finds of rice grains from caves in Borneo and South Sulawesi suggest an early date for the emergence of rice cultivation in the region but, at present, a precise dating for the substantial quantity of carbonised, domesticated rice grains and chaff from Ulu Leang I cave in South Sulawesi remains unresolved. Indeed, our understanding of the introduction of agriculture into Indonesia generally, is hindered by the lack of well reported excavations and surveys. Much of the information available on the history of rice has been gleaned from linguistic analyses and from items of material culture, such as pottery and polished stone tools. Although more reliably dated materials are urgently needed, it appears that once rice cultivation became established in Indonesia it spread very rapidly throughout the region.

Migrants from the North

Evidence suggests that Austronesian-speaking horticulturalists first began moving into the Indonesian region, from a homeland in southeastern China and Taiwan, some 6,000 years ago. They came via the Philippines and, as they headed south and westwards, they absorbed and demographically swamped the sparsely scattered populations of hunter-gatherers who had previously settled in the

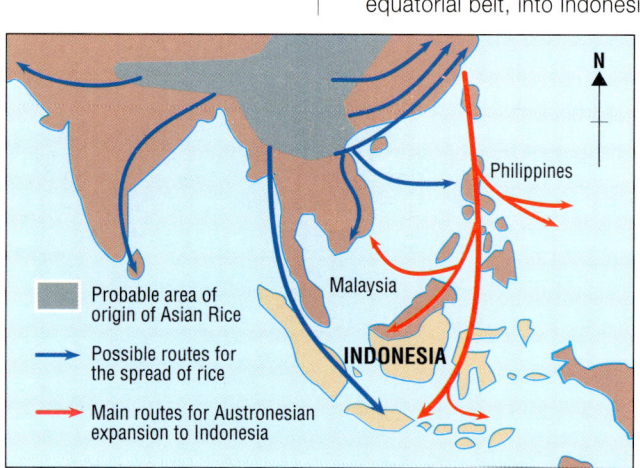

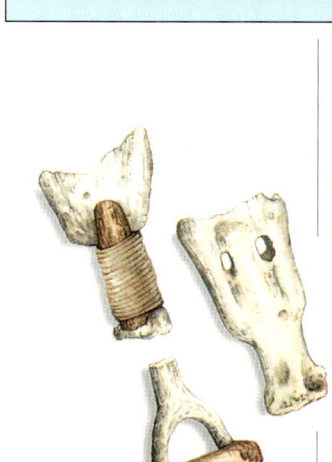

The remains of 6,400-year-old spades used when rice was first cultivated in Southern China.

area. The latter probably spoke languages related to those found in present-day New Guinea in the east, and parts of mainland Southeast Asia. Much of the evidence for these conclusions comes from artefacts discarded or lost in caves and from piles of marine shells which occur at former settlements in Sumatra, Java, Sulawesi, Timor, East Malaysia and the Philippines. The items include pottery and polished stone tools; bones and plant remains have been collected but only occasionally identified.

As Austronesian speakers expanded eastwards into Melanesia and the western Pacific, these people gradually replaced their cereal crops, which probably included foxtail millet as well as rice, with tubers and tree fruits. The western branch of Austronesian speakers, however, retained rice as a significant crop plant as they moved into Sulawesi, Borneo, Java and Sumatra. This subsequently developed an intensive system of wet rice cultivation, with terraced rice fields, following the introduction of iron, domestic buffaloes and cattle, some 2,000 years ago.

Other Early Cultivated Plants

Other economically important domesticated plants native to Indonesia include millet (*Setaria italica*), and root crops such as taro (*Colocasia esculenta*), and yams (*Dioscorea spp.*). Native fruit trees such as coconut, mango, durian, mangosteen, breadfruit and jackfruit, as well as vegetables, have also been cultivated for several thousand years. Many other cultivated plants which are common and widespread today, such as maize, cassava, peppers and tomatoes are of post-colonial, American origin.

Cultivation Systems

There are two main types of agriculture traditionally practiced in Indonesia. The first is shifting cultivation, which is low-intensity with long fallow periods; it is commonly found in the hills. Hill rice is the typical cultivar, but many other cereal, fruit and vegetable species are also grown. In the lowlands, where there is greater soil fertility, there is a tendency towards intensive, fixed-field systems growing one major species. The latter is mainly represented by the paddy fields of wet rice cultivation, but one occasionally finds intensive root and tuber cultivation in high mountain valleys, particularly in Irian Jaya.

Eastern Evidence

There is little reliable evidence by which to date the first appearance of agriculture east of Timor, until one reaches eastern Melanesia where the proto-Polynesian 'Lapita Culture' was established by about 4,000 years ago. The economy of lapita folk was very much adapted to fishing and coastal life, but they almost certainly cultivated yams, breadfruit, candlenut and the canarium tree. They are also thought to have kept some dogs and pigs. In the New Guinea highlands, however, there is good evidence of a much earlier, and most probably independent horticultural tradition. Drainage ditches, thought to be for taro and other root crop cultivation, have been dated back to 9,000 years ago.

The seed heads of foxtail millet and its likely wild ancestor, green bristle-grass, demonstrate an increase in head size with domestication.

Dayak women working in the hills of the upper Mahakam in Kalimantan, form a sharp contrast with farmers transplanting rice in the wet paddy fields of the lowlands of Sulawesi.

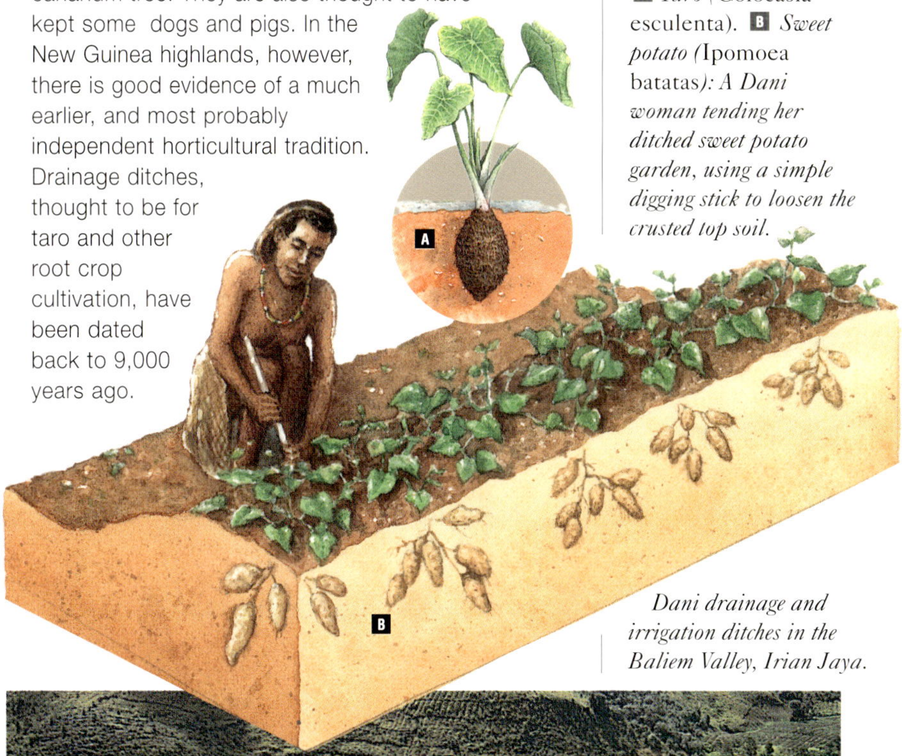

Crops with a preference for wet and dry environment respectively: **A** *Taro (*Colocasia esculenta*).* **B** *Sweet potato (*Ipomoea batatas*): A Dani woman tending her ditched sweet potato garden, using a simple digging stick to loosen the crusted top soil.*

Dani drainage and irrigation ditches in the Baliem Valley, Irian Jaya.

Rice and Rice Fields

Rice is one of the most nourishing and delicious staple foods. A large population can be supported by quite a small rice growing area which produces good harvests year after year. There are costs, however: farmers have to work together to share water supplies and intensive labour is required to maintain the small fields and watercourses. A storm, volcanic eruption, a plague of rats or the outbreak of a new disease can destroy a crop in a few days or even hours.

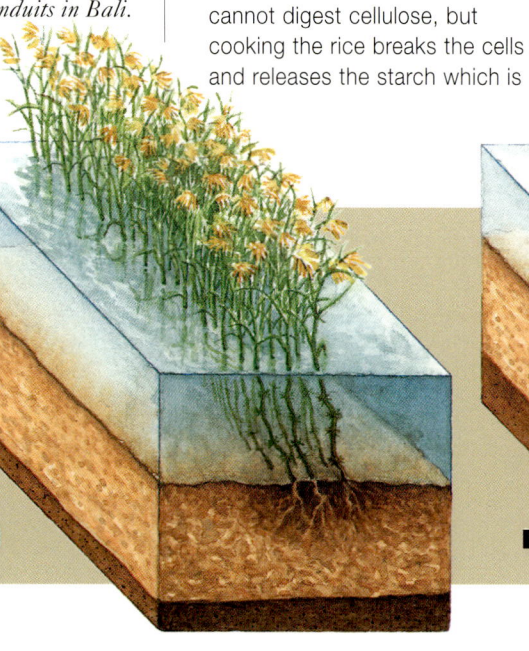

HARVESTED AREA OF RICE BY ECOSYTEM (,000 HECTARES)

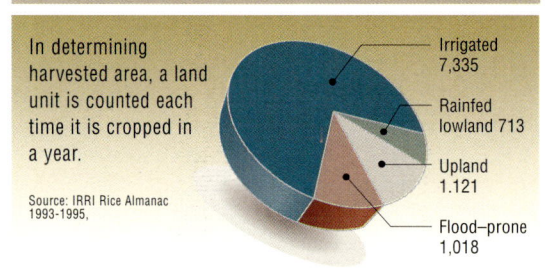

In determining harvested area, a land unit is counted each time it is cropped in a year.

Source: IRRI Rice Almanac 1993-1995.

- Irrigated 7,335
- Rainfed lowland 713
- Upland 1,121
- Flood-prone 1,018

Cutting paddy in West Java, using an ani-ani, a small knife, half hidden in the palm so that the rice soul is not frightened or offended.

☘ *Holy water from Gunung Agung irrigates the field via an organised system of bamboo conduits in Bali.*

Rice Plants and Ecosystems

Rice is the name given to a number of grasses, the most important genus of which is *Oryza* with some 24 species. Only two of these — *O. sativa* and *O. glaberrima* — are cultivated, and the latter only in West Africa. The varieties of *O. sativa* are almost unlimited because it is easy to form hybrids. The development of genetic science has resulted in their rapid breeding, 'tailored' to their ecosystems. Seeds of about 80,000 varieties are stored in the germ-plasm bank at the International Rice Research Institute (IRRI) at Los Baños, in the Philippines.

There are four major rice plant ecosystems: dry upland rice, grown in the forested hills of Kalimantan; rainfed rice, found in wet, lowland areas without irrigation; irrigated rice, which is about three-quarters of the rice grown in Indonesia; and flooded or deep water rice, with five metres high stems to keep seeds and flowers above the surface of flooded lakes, grown in parts of central Java and West Kalimantan.

Rice Grains

The seed of the rice plant stores energy for the embryo and young plant during germination and the first few days of life. This energy is in the form of starch found in tiny cells with cellulose walls. Humans cannot digest cellulose, but cooking the rice breaks the cells and releases the starch which is converted to glucose by our bodies. The cellulose acts as fibre, helping the body get rid of wastes.

The rice grain has a brownish outer coat of bran containing vitamins, proteins and minerals for the young plant. Most rice is milled to make it white, and the bran coat is lost (much is used for animal feed). However, white rice still contains some protein, B-group vitamins and minerals.

One of the soft starches is amylose: the more there is of it, the less sticky the cooked rice will be. Indonesians like *javanica*, the long-grained, soft and sticky rice. In northern India and the west, *indica*, with its long grains remaining separate on cooking, are preferred, while the sticky, short-grained variety, *japonica*, is popular in East Asia.

Green Revolution — Benefits and Costs

The biggest yields, and the best-tasting rice, comes from irrigated fields which can be flooded with stored water and drained again. Sumatra, Java, Bali, parts of Sulawesi, and some of the smaller Indonesian islands have a high rainfall, hours of sunshine and fertile soils, made more so by volcanic minerals.

Although rice production has increased greatly during the past decade, the yields have not kept pace with the demands of a growing population. By the early 1970s, a possible severe food shortage

PEOPLE AND PLANTS: CULTIVATION

EXAMPLES OF ORGANISMS THAT MAY HARM OR COMPETE WITH THE RICE CROP

INSECTS			DISEASES OF RICE	
Stem borers			**Viral diseases and their vectors**	
Yellow stem borer	*Scipophaga incertulas* (Walker)		Rice tungro	*Nephotettix virescens* (Distant)
White stem borer	*Scipophaga innotata* (Walker)			*N. nigropictus* (Stål.)
Striped stem borer	*Chilo suppressalis* (Walker)		Ragged stunt	*Nilaparvata lugens* (Stål.)
Dark-headed rice	*Chilo polychrysus* (Meyrick)			*L. acuta* (Thunberg)
Defoliators			**Bacterial diseases and their causal agents**	
Rice leaffolders	*Cynaphalocrocis medinails* (Guenée) and others		Bacterial blight	*Xanthomonas oryzae* pv. *oryzae* (Uyeda ex Ishiyama 1922)
Rice caseworm	*Nymphula depunctalis* (Guenée)		Whitebacked	Swings et al 1990
Leafhoppers			**Fungal diseases and their causal agents**	
Green leafhopper	*Nephotettix virescens* (Distant)		Blast	*Pyricularia oryzae* Cav.
	N. nigropictus (Stål.)		Sheath bright	*Rhizoctonia solani* (Thanatephorus cucumeris) [Frank] Donk
	N. parvus Ishihara et Kawase			
	N. cincticeps (Uhler)			
Planthoppers	(not illustrated)		**RODENTS**	
Brown planthopper	*Nilaparvata lugens* (Stål.)		Rice field rats	*Rattus rattus argentiventer* (Rob & Kloss)
Whitebacked planthopper	*Sogatella Furcifera* Horvath			*R. r. Mindanensis* (Mearns)
Rice bugs				
Malayan black rice bug	*Scotinophara coarctata* (Fabricius)			
Rice grain bug	*Leptocorisa oratorius* (Fabricus)			

Source: IRRI Rice Almanac: 1993-1995.

was averted by a 'green revolution', with its high-yielding varieties (HYVs), programmes for irrigation and efficient distribution of food and fertilisers.

The first HYVs were susceptible to attack by pests and diseases and required fertilisers which made them expensive to grow. Besides, the taste and texture of the cooked rice was unsatisfactory. Most of these problems have now been solved, but HYVs bring new demands. The new rice can only be cultivated from specially grown seed, which must be bought by the farmer, together with chemical fertilisers, if their maximum yield is to be realized. Traditional rice varieties are often not allowed where HYVs are grown, resulting in the loss of many old rice varieties highly valued for their flavour, texture and resistance to pests and diseases.

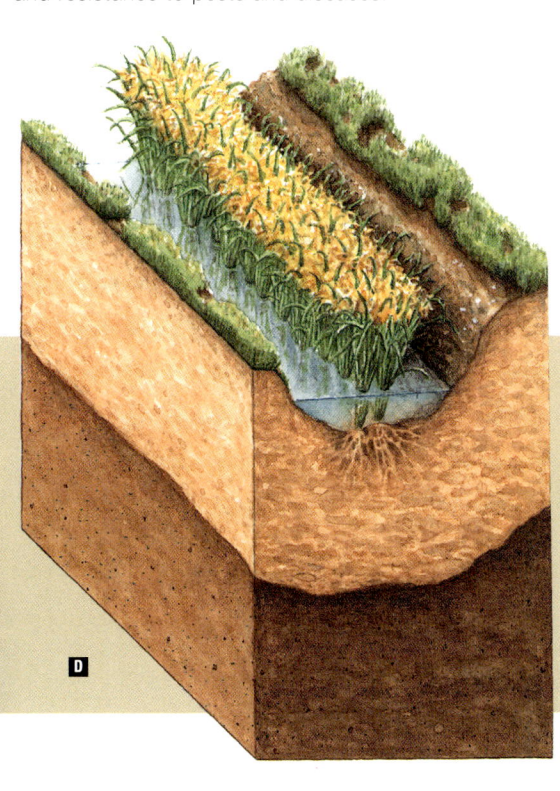

According to IRRI, between 1990 and 2020, the rice yield per hectare must increase by 60 per cent to feed the world's population. Indonesia is now able to grow enough rice to feed its current population, at prices which are fair and stable to producers and consumers alike.

THE DIFFERENT ECOSYSTEMS IN WHICH RICE IS GROWN

The ecosystems are characterised by various agroecological factors such as elevation, rainfall pattern, depth of flooding and drainage and also the adaptability of rice.

These are as follows:
- **A** Deep water
- **B** Tidal wetland
- **C** Rainfed lowland
- **D** Irrigated
- **E** Upland

In the flood-prone ecosystems, that is deepwater and tidal wetlands fields are characterised by medium to very deep flooding (50 to more than 300 centimetres) from rivers and tide in river mouth deltas. Slightly sloping, bunded or dyked fields with variable depth and duration of flooding are features of rainfed lowland. On the other hand, the upland is characterised by flat non-flooded, well-drained soil from level to steeply sloping fields.

Indigenous Gardens: Re-Inventing the Forest

Indonesian agriculture is rooted firmly in the forest, with native gardens, or kebun, representing an intermediary step between the simple exploitation of forest resources and the development and management of conventional agricultural systems. These gardens are forms of 'agroforestry', a term which brings together the natural forest and agriculture through the cultivation of wild forest tree species.

Over a century ago, farmers on the west coast of Lampung in southern Sumatra were traditionally resin collectors. They now integrate the best resin producing species in their subsistence gardens, covering thousands of hectares.

A home garden near Tanjung Pinang in Pulau Bintan, the largest town in the Riau Archipelago, Sumatra. Bananas and sugarcane can be seen growing in the foreground.

Traditional Subsistence Gardens

The most primitive form of indigenous gardens are 'fruit tree islands' which are found scattered in the fallow lands of shifting agriculturalists, or at traditional resting places along established pathways. These are where people would have eaten and spat out the seeds of the fruit they had brought with them for their journey. These subsequently evolved into more complex garden forms which today surround villages all over Indonesia.

The composition of these subsistence gardens typically includes domestic plant species, such as vegetables, fruit, coffee, coconut and rubber, as well as wild or semi-wild forest species. The sophisticated 'home gardens' of Java, where a large variety of species are grown for a wide range of purposes, constitute a highly evolved version of the traditional subsistence garden in a densely populated region.

Tapping the Forest

The ancient trade in forest produce has, since early times, influenced the composition and development of Indonesia's subsistence gardens and encouraged the incorporation of many forest species into the garden system. In Sumatra, there is evidence that by the 16th century, valuable cinnamon trees (*Cinnamomum burmanni*) and benzoin (*Styrax benzoin*) were being brought out of the forest and planted close to settlements as the demand for their products grew.

Cash Cropping

Native gardens often cover large areas between villages and rice fields or swidden fields cleared in the forest, and may include several commercial species as well as purely subsistence crops. In many instances they represent the sole source of family income for the household concerned. During the colonial era, native gardens played an important role in the establishment and extension of a

Forest farmers weighing blocks of latex prior to selling them.

THE KEBUN

The typical structure of native gardens or *kebun* (right) all over Indonesia includes a high canopy of fruit, nut or timber trees, with the lower tree communities comprising commercial species such as cinnamon, nutmeg, coffee, rubber, or fruit species and natural forest treelets. Bamboos and palms are also important plants in these situations. The ground is generally covered with herb species characteristic of forest undergrowth such as ferns, begonias, aroids and gingers.

The drying of the cinnamon tree bark in Sumatra, which is used as a spice.

FLORISTIC COMPOSITION OF AN AGROFORESTERY PLOT (1000 sq. m) IN MANINJAU, WEST SUMATRA (TREES DBH 10 cm)

Scientific name	Family	No.	Type
Durio zibethinus	Bombac.	15	fruit/timber
Pterospermum javanicum	Sterculiac.	23	timber
Alangium kurzii	Alangiac.	3	timber
Cinnamomum burmanii	Laurac.	85	spice (bark)
Myristica fragrans	Myristic	4	spice (nut)
Pandanus tectorius	Pandanac.	1	material (leaves)
Schizostachyum brachycladum	Poac.	2	material (culms)
Toona sinensis	Meliac	1	timber
Eugenia malaccensis	Myrtac.	1	fruit/timber
Bridelia monoica	Euphorbiac.	1	greens (leaves)
Baccaurea dulcis	Euphorbiac.	1	fruit/timber

Source: After Genevieve Michon and Hubert de Foresta, 1990.

market economy based on agricultural production: the real 'rubber invasion' at the beginning of the 20th century, occurred in jungle smallholdings in the lowlands of Sumatra and Kalimantan where rubber trees were cultivated alongside numerous indigenous species. Even today, when rubber cultivation is associated with enormous, uniform plantations, there are still 2.5 million hectares of rubber under cultivation in gardens in the eastern lowlands of Sumatra. Indeed, rubber small holdings account for more than half of Indonesia's total export of rubber.

But if native gardens assisted the establishment of imported species in Indonesia, they have equally played a major role in introducing indigenous species into the world market economy. In this respect, the traditional native garden has played an important part in integrating an indigenous system of agricultural production into the modern economy.

A New Alliance

Many cultures make a fundamental economic distinction between forest and agricultural land — the wild and the domesticated. Yet, this is not the view shared by many Indonesian farmers, however, who in their long acquaintance with the forest have learned how to maintain forest-like gardens. Even in areas which are located at a distance from natural forests, it is noticed that villagers have still managed to incorporate forest species in their gardens and continue to develop a system of mixed planting and cropping. The experience, which must have been garnered over centuries, through trial and error, has equipped native farmers all over the Archipelago with a unique understanding of agroforestry. This, in turn, could prove invaluable to the development and progress of both agriculture and forestry in the region.

STOREHOUSES OF BIODIVERSITY

Indigenous gardens have a very high biological diversity, and their conservation represents a major contribution to development by farmers. These gardens not only allow the direct conservation of numerous useful forest species, but also act as shelters for hundreds of wild forest species not directly useful to people. Near Padang, West Sumatra, one of the richest and most densely populated provinces outside Java and Bali, natural lowland forests have long ago been cleared for permanent agriculture, but indigenous gardens, in which more than 150 tree species may be managed, allow the conservation of several hundred natural forest species, from orchids to timber trees, and from palms to lianas. Wild animals found in these forests include monkeys, squirrels, bats and civets cats, and if the gardens are adjacent to natural forests they may receive visits from gibbons, bears, small cats, and even tigers. In addition to being storehouses of biodiversity, these gardens also represent invaluable reservoirs of information on useful plants, including local uses, ecology, and the responses of complex plant communities to different forms of human management. Such local contributions to conservation should not be overlooked. Early subsistence gardens like this one (above) in East Kalimantan are established around resources traditionally sought after in the natural forest for subsistence purposes: food species such as fruit and spices; fibre species for ropes and clothes; rattans, bamboos and palms for handicraft and building purposes and all kinds of shrubs; while lianas and herbs are put to various uses.

The massive ceiba or kapok tree (Ceiba pentandra), bearing large pods filled with seeds invested with a silky floss. It is a common village tree in Sulawesi. The silky fibres are used as filling for mattresses, cushions, sleeping bags, life preservers, and as insulation.

Rambutan (*Nephelium lappaceum*) sp.)
Timber (*Pterospermum javanicum*)
Nutmeg (*Myristica fragrans*)
Cinnamon (*Cinnamomum burmanii*)
Durian (*Durio zibethninus*)
Bamboo (*Schizostachyum branchycladum*)
Fruit tree (*Baccaurea sp.*)
Fruit Tree/Timber (*Toona sinensis*)
Timber (*Alangium kurzii*)

Source: After Wiyono, 1996.

Legumes: An Introduction

An estimated 18,000 species of legumes have been recognised worldwide, making them one of the largest families in the plant kingdom. In Indonesia, there are approximately 510 native species of legumes grouped in 110 genera. Many of these plants provide valuable services to people, whether as ornamental species, timber trees, medicinal and poisonous plants, green manure, ground cover, or as a source of food.

Bright, fiery red flowers of the exotic ornamental, Delonix regia *were introduced from Madagascar.*

What Makes a Legume?

The legume family (Leguminosae) takes its name from the Latin *legumen,* or pod. This is a two-valved fruit with seeds attached to the edge of the upper margin. The pod may be straight, coiled, rounded, flattened, winged, woody, fleshy, smooth or hairy. When ripe, they may split open along the margin or sometimes break into seeded segments. Many seeds are pea- or bean-shaped; all have two cotyledons and an embryo, and can often retain the capacity to germinate for many years.

Pods of different species can display a number of interesting features. The pod of *Entada rheedei* is the largest fruit in the world and may grow up to one metre in length. This plant is a large liana, but its mimosoid flowers only measure one centimetre each. On hot, dry days the pods of some species, for example *Bauhinia,* snap open with an audible crack, ejecting seeds several metres into the air.

Legume Taxonomy

Taxonomists divide the legume family into three sub-families. The Papilionoideae or butterfly-flowered legumes which have fused stamens, cover most species (12,000) and include most food legumes. The Caesalpinioideae (3,000 species) also have two-sided or zygomorph flowers, that is, flowers possessing only one plane of symmetry. The petals and stamens do not grow together as closely as they do in papilionoids. This sub-family contains important timber and ornamental species. The third group, the Mimosoideae (3,000 species), include the *Acacia* and *Mimosa* species which possess beautiful floral pompoms with stamens protruding from the rest of the flower. All members of these sub-families produce a seed pod, but these may vary a great deal in size and shape.

Representatives of the legume sub-families Caesalpinioideae, Mimosoideae *and* Papilionoideae *(from left):* Caesalpinia pulcherrima, Acacia farnesiana *and* Pisum sativum *(the pea).*

The Papilionoid Flower

All papilionoid flowers are shaped according to the same plan, although sometimes they grow upside down! The five petals can be very showy, and attract insects to the nectar at their base. Insects land on the keel, still covering the staminal tube, guided by lines on the 'flag' petal which may be visible only in ultra-violet light. Some species have explosive stamens that eject from the keel when insects try to enter, thereby releasing the pollen grains. Cross-pollination is common, but so is self-pollination; the stigma then grows through the anthers on the staminal tube whilst still enveloped in the keel, and gathers pollen from the anthers of the same flower.

Versatile Plants

Herbaceous legumes are a useful component of grasslands because they add beneficial nitrogen to the soil. In addition to helping control soil erosion, shrub and tree legumes provide shade, wind cover and green manure for the soil, as well as mulch for plantations and vegetable gardens to keep the soil covered and cool. Hedges of legumes provide demarcation and cover in fields, providing protection for crops. Important cover plants like tropical kudzu (*Pueraria phaseoloides*), calopo (*Calopogonium mucunoides*), and centro (*Centrosema pubescens*) are extremely useful in oil palm and rubber plantations. Apart from their nitrogen-fixing ability, they are also good phosphate miners. When plantation trees are young, cover plants grow rapidly across the soil, protecting it from erosion, and providing appropriate habitat for the soil organisms

PEOPLE AND PLANTS: CULTIVATION

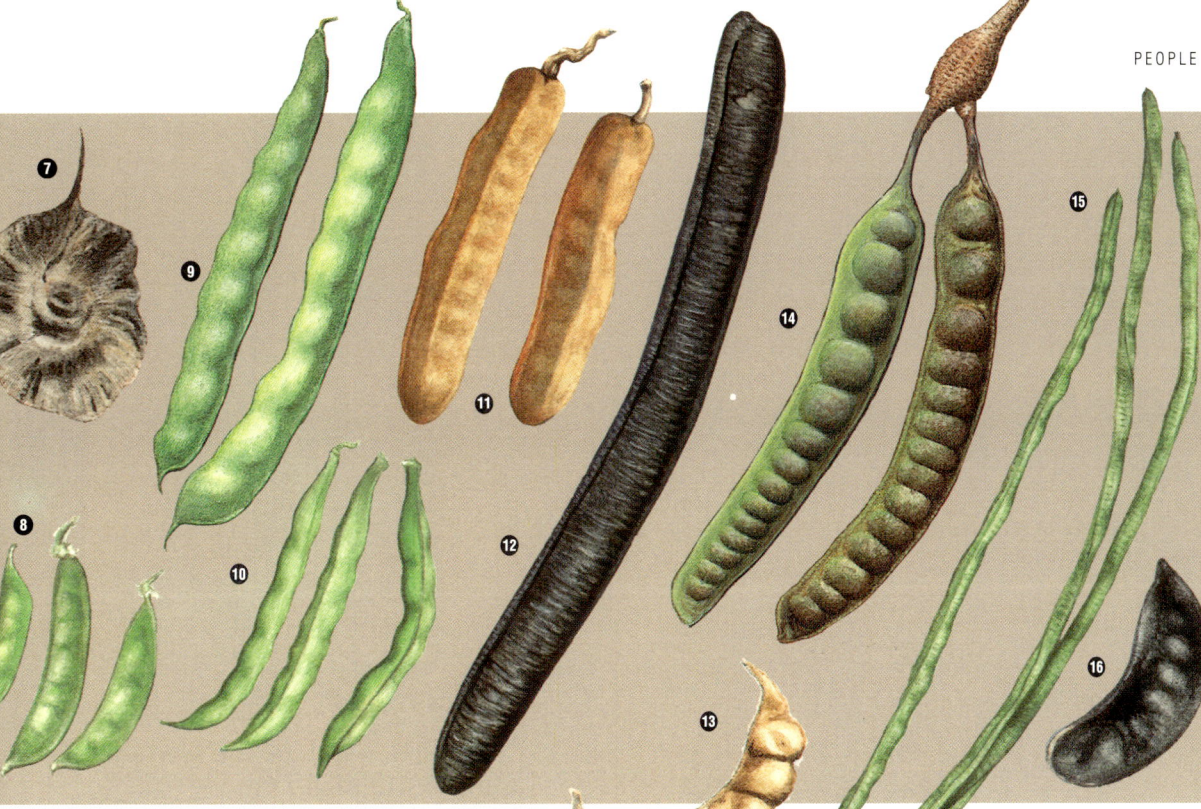

AN ARRAY OF PODS OF DIFFERENT SHADES, SHAPES AND SIZES

1. *Entada rheedei* Sprengel
2. *Pterocarpus rohrii* Vahl
3. *Ornithopus sativus* Brot.
4. *Vigna subterranea* (L.) Verdc.
5. *Cicer arietinum* L.
6. *Mimosa pigra* L.
7. *Pterocarpus indica* Wild.
8. *Pisum sativum* L.
9. *Phaseolus vulgaris* L.(besar)
10. *Phaseolus vulgaris* L.
11. *Tamarindus indica* L.
12. *Mimosa invisa*
13. *Cajanus cajan* (L.) Millsp.
14. *Parkia speciosa*
15. *Vigna unguiculata* (L.) Walp.
16. *Senna alexandrina* Miller

needed for the decomposition of organic matter. They can also smother and kill unwanted weeds. Most cover legumes are also palatable to cattle.

Valuable Timber Species

Many valuable timbers are legumes, including 42 of the major commercial timber trees in Indonesia. The major leguminous timbers grown in Indonesia are Indian rosewood (*Dalbergia*), keranji (*Dialium*), merbau (*Intsia*), kempas (*Koompassia*), batai or sengon (*Paraserianthes*), nandu wood (*Pericopsis*), red sandalwood (*Pterocarpus*), and sepetir (*Sindora*). Each genus has one or more species with good timber qualities. These tropical hardwoods and luxury timbers earn foreign exchange, while local importance for construction and making furniture is even greater.

Ornamental Species

Well-known and frequently planted ornamental legumes include the flamboyant tree (*Delonix regia*) from Madagascar, which today adorns many roadsides in Indonesia with its masses of red flowers which appear around Christmas time. The Burmese *Amherstia noblis* is another beautiful species and is often planted in botanic gardens because of its long, hanging inflorescences of red flowers. The young leaves are pinkish and also hang like tassles from the branches. This 'flush' is typical of many tropical trees, where soft new growth of yellow, white, light green or purple-coloured leaves develops. After a showy period of a week or so, the leaf chlorophyll takes over and the leaves unfold, adopting their usual green colouring. Small trees, such as the orchid flower tree (*Bauhinia purpurea*), with its purple and white flowers, are easily grown. The genus *Erythrina* (dadap) has splendid scarlet or red flowers. The Asian coral tree (*Erythrina variegata*) has red- or white-flowered cultivars, and some have very decorative yellow bands on the main veins of the leaflets. One species, *E. orientalis* grows up to 15 metres, while the Brazilian *E. crista-galli* is a very handsome shrub which may reach a height of three metres.

Ornamental seeds are produced by the shade and timber tree *Adenanthera pavonina*, or the climber *Abrus precatorius*: they show a vivid contrast of black and red. *Saracca* flower bunches have bright orange colours. The distinctive rain tree (*Samanea saman*) has pinkish and white powder-puff flower bunches, and folds its leaflets before sunset (*pukul lima*) and when it rains.

The rain tree (Albizia saman) *is a handsome spreading shade tree common in central squares of many Indonesian towns. It originates from tropical America.*

«« *The flame of Irian* (Mucuna bennettii) *is one of Indonesia's most spectacular ornamentals.*

LEGUME LEAVES
Legume leaves have some common features although their shape and size can be very different. Virtually all leaves are compound, even those that appear to be simple! This is shown by the two swollen 'elbow' joints at both ends of the petiole which are absent in simple leaves. The trifoliolate leaves of many beans are well known, but legume leaves can be once or twice compound, the latter being the rule in the Caesalpinioideae and Mimosoideae sub-families. Simple, few-foliolated and simply pinnate leaves are common in the Papilionoideae.

Legumes: Food and Forage

Rich in protein and oil, legumes such as soyabeans, yard-long beans and peanuts are wholesome additions to our daily diet, complementing the main carbohydrate staples of cereals such as rice and maize. In addition, leguminous trees, shrubs and herbs play an important role as fodder species for livestock.

»»*A cluster of Petai or Stinking Beans (*Parkia speciosa*) pods, a popular vegetable and important commercial crop of Indonesia. Its main areas of cultivation include West Java, Central Java and Sumatra.*

Edible Legumes

The legume family owes its protein richness to a symbiotic relationship with root-nodule bacteria which fix free nitrogen from the air. Pulses or grain legumes are those legume seeds eaten after ripening. Preparation includes soaking before cooking, or sprouting. Pulses can be stored dry for a long time if protected against bruchid beetles and weevils.

The soyabean (*Glycine max*) is the most important oil-supplying crop in the world, but in Asia a large variety of fresh, fermented and dried food products are made from it. The other major oilseed is groundnut or peanut. Together they provide salad oil, cooking oil, margarine and shortening (oil palm and coconut are also important in this capacity in the wet tropics). The protein-rich cake remaining after extraction is valuable as cattle feed.

One of the world's most popular snacks, the South American peanut or groundnut (*Arachis hypogaea*) was first introduced to Indonesia from Brazil and Mexico. Today, this legume is cultivated more in Southeast Asia than in its native countries, and comprises about 10 per cent of the world production. A major world source of oil and food, the woody pods of this legume are borne on stalk-like pegs which develop in the soil. After the flowers, which are carried in the leaf axils, are fertilised, they grow downwards into the soil. This peculiar habit is called geocarpy, and known only in a few legume species. Bambara groundnut or Kacang Bogor (*Vigna subterranea*) is another example. Groundnut pods have one to four, sometimes up to six seeds, and the oil- and protein-rich kernels are used for oil, peanut butter, snacks, and even cooked as vegetables. Kacang Bogor is lower in fat, and has only one seed per pod, that is used as snack or as vegetable.

Another vegetable legume produces industrial vegetable gums: guar (*Cyamposis tetragonoloba*) seeds contain

*The sour tasting pulp from the indehiscent pod of Tamarind (*Tamarindus indica*) is made into paste, a widely used component of Indonesian cuisine.*

*Baskets filled with processed soyabean (*Glycine max*), in West Java, ready to be milled and made into 'Tahu' or soyabean curd.*

galactomannan, with a thickening property up to eight times stronger than starch. It is used as a stiffener and stabilizer in various food items. Guar pods are also eaten as vegetable or snack, and the plant is used for green manure and cover or shade crop.

Vegetables and Spices

Many legumes are used as vegetables (the French word légume means vegetable) and Indonesia produces both tropical and temperate kinds, although the latter are grown only in mountain gardens. The yard-long bean (*Vigna unguiculata*) and the mungbean (*V. radiata*), which is world famous for its bean sprouts, are both of Asian origin. Green beans (*Phaseolus vulgaris*), came originally from America but are now commonly grown throughout the world. *Bengkowang* (*Pachyrhizus erosus*), with its sweet tubers which are eaten raw in *rujak* fruit salad, is also from America. Even the flowers of leguminous plants may be eaten as vegetables as, for example, in the case of *Sesbania grandiflora*, which is probably the world's largest legume flower. More a spice than a vegetable, the strong-tasting seeds of *petai* (*Parkia speciosa*) are used in various culinary dishes.

The tamarind tree, or *asam jawa* (*Tamarindus indica*), is a native of tropical Africa but is cultivated in Indonesia for its edible fruit. These take the form of large plump pods containing flat seeds embedded in a soft, brownish flesh. The flesh is an important ingredient in Indonesian cuisine, most notably in the preparation of the sauce for rendang dishes.

Legumes as Fodder

Leguminous trees, shrubs and herbs play an important role as fodder for foraging species. Pastures are much improved by fast-growing legume creepers. These may be naturally occurring or introduced species. In agroforestry, legume shrubs and trees can be lopped or browsed for forage. Good examples include the Central and South American plants *Leucaena leucocephala*, *Gliricidia sepium* and *Calliandra calothyrsus*. The latter has become especially popular in Indonesia as a fodder tree which has an additional ornamental value. These species may also be used for fuel, as fire or wind breaks, and form attractive living fences.

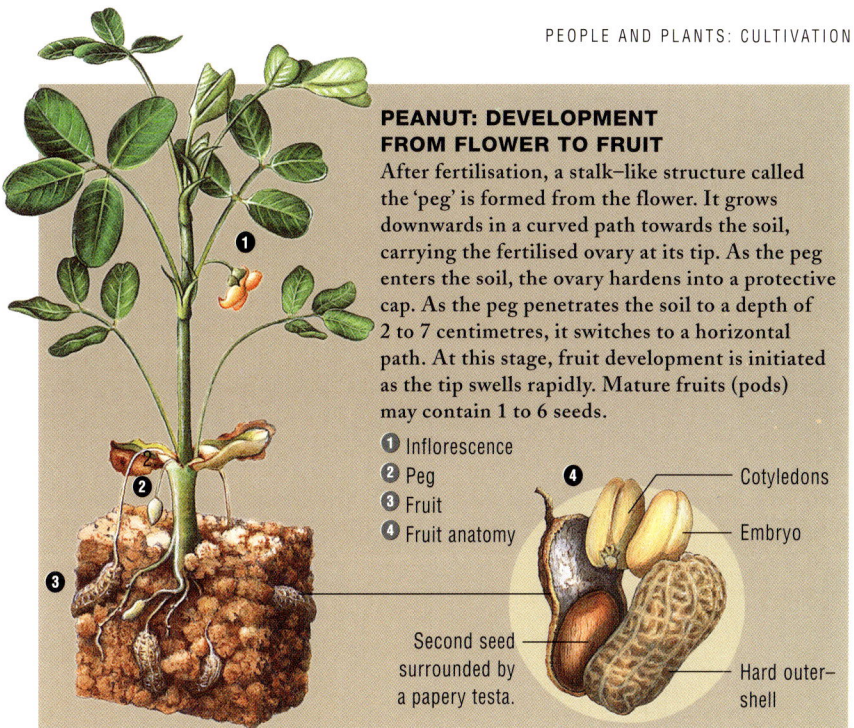

Too Many Beans

An unpleasant, but quite natural, aftereffect of eating too many pulses, is wind produced from the bowels. Two sugar components in pulses, stachyose and verbascose, are difficult to digest as no enzymes can break them down. As a result, intestinal bacteria ferment the pulses, resulting in abnormal level of gases. The accompanying odour of fermentation is caused by methane and other volatile compounds. Well-prepared legumes in a balanced diet, eaten regularly, cause no undue trouble. Some people, however, have more sensitive bowels than others.

PEANUT: DEVELOPMENT FROM FLOWER TO FRUIT

After fertilisation, a stalk-like structure called the 'peg' is formed from the flower. It grows downwards in a curved path towards the soil, carrying the fertilised ovary at its tip. As the peg enters the soil, the ovary hardens into a protective cap. As the peg penetrates the soil to a depth of 2 to 7 centimetres, it switches to a horizontal path. At this stage, fruit development is initiated as the tip swells rapidly. Mature fruits (pods) may contain 1 to 6 seeds.

1. Inflorescence
2. Peg
3. Fruit
4. Fruit anatomy

A street vendor with her wares: peanuts, a simple but versatile food legume which may be eaten in its original form or processed into other food products.

THE WONDER LEGUME

All parts of the climbing winged bean (*Psophocarpus tetragonolobus*) from Burma and the island of New Guinea can be eaten. The young leaves and flowers may be served as a salad and the pods as vegetables (their most popular use); the seeds can be cooked as a pulse, used as oilseed or fermented into cakes; while the tubers produce an edible starch. Some scientists have called it 'a supermarket on a stem', but internationally-funded research has not delivered the expected benefits of this undoubtedly exceptional crop.

The plant is still largely semi-domesticated, and future breeding trials have to suppress some of its primitive habits, such as climbing, seed scattering and a perennial nature. These characteristics are found in many ancestors of our legume crops but domestication has produced annual, self-supporting genotypes which retain their seeds.

A. Section across a mature fruit which may contain 5 to 20 seeds.
B. Flowering and fruiting branch
1. Trifoliate leaves
2. Flowers
3. The developing fruit can grow from 6 to 40 centimetres long and 2.5 to 3.5 centimetres wide.

Root and Tuber Crops

Root and tuber crops have been of importance throughout the history of the Indonesian Archipelago. Many of the currently important species have been introduced. They are often a major food source in relatively isolated areas and are also of some considerable importance in modern food processing industries. Root crops are therefore of continuing economic value in Indonesia

*Farmers selling freshly harvested cassava or tapioca (*Manihot esculenta*) in the local market.*

*Work in a sweet potato (*Ipomoea batatas*) field in Bali during harvesting — digging for the root crop, collecting it into baskets, and carrying basketful of tubers out of the fields.*

»»*Dani women steambaking the harvested sweet potatoes for a celebration in Milinggame, Tioru, Irian Jaya.*

ROOTS AND TUBERS IN INDONESIA

MAJOR CROPS

Common Names	Latin Names	Family
Cassava	Manihot esculenta	Euphobiaceae
Sweet potato	Ipomoea batatas	Convolvulaceae
Potato	Solanum tuberosum	Solanaceae
Taro (Cocoyams)	Colocasia esculenta	Araceae
Greater yam	Dioscorea alata	Dioscoreaceae
Kerung	D. nummularia	Dioscoreaceae
Ubi pasir	D. pentaphylla	Dioscoreaceae
Asiatic yam	D. esculenta	Dioscoreaceae
Potato yam	D. bulbifera	Dioscoreaceae

LESS IMPORTANT CROPS

Common Names	Latin Names	Family
Arrowroot	Maranta arudinaceae	Marantaceae
Tannia (Cocoyams)	Xanthosoma sagittifolium	Araceae
Yam bean	Pachyrhizus erosus	Leguminosae
Waternut	Eleocharis dulcis	Cyperaceae
Chinese potato	Solenostemun rotundifoius	Labiatae
Edible canna	Canna edulis	Cannaceae

Industrial and Domestic Uses

A large number of roots and tubers are grown in Indonesia — cassava, sweet potato, taro, yam and potato — all of which have numerous varieties. Cassava is a major crop in the hilly areas of Java, where consumption reaches over 60 kilogrammes per capita per year. In the eastern islands, and especially in Irian Jaya, sweet potato, taro and yam form the dietary staple. The starch laden roots of these crops are not the only valuable parts of these plants: cassava leaves are eaten as a nutritious vegetable, while the leaves of the sweet potato provide fodder for animals.

Today, cassava is a major raw material in Indonesian industry, with starch from cassava being used in the wood and textile industries and in the production of snacks and processed foods. This starch industry is centred in East and Central Java and in Lampung, South Sumatra. Cassava factories also make chips and pellets for animal feed.

Cultivation of Root Crops

Root crops can be successfully cultivated under a wide range of agricultural conditions. While cassava is grown in the uplands on non–irrigated land, sweet potato may be cultivated on both non–irrigated and rain–fed irrigated land. Cassava takes around seven months to mature, while sweet potato and potato crops are ready for harvesting after four or five months. Cassava is propagated simply by planting the stem in the ground, and sweet potatoes by planting the runners. In both cases, propagation is easy and does not affect the major usable part of the plant. Cassava root is highly perishable once harvested, but keeps well while in the ground. This means that farmers only need to harvest roots as and when they require them. Mature sweet potato and potatoes, on the other hand, can be kept for some time if stored at ambient temperatures. Farmers can thus arrange an almost year–round supply of food from root crops.

Origins of Major Root and Tuber Crops

Cassava has its origins in the northern part of South America, while the sweet potato and potato come from the Andes Mountains. Cassava had been introduced into Indonesia by the late 18th century with the potato arriving around the same time. Sweet potato, taro and other root crops were already present when the Dutch arrived in eastern Indonesia in 1618, but virtually nothing is known about their introduction to the Indonesian Archipelago.

Irian Jaya's Sweet Potatoes

The sweet potato contributes about 80 per cent of the calorific intake in most highland societies of Irian Jaya. Daily consumption ranges from 1.5 to three kilogrammes, depending on age, gender and local dietary habits. The importance of these in local subsistence agriculture is reflected in the hundreds of local cultivars. Botanists and geneticists consider the island of New Guinea as a secondary centre of diversity for these plants: although the crop does not originate from the area, the rugged terrain, together with human selection and propagation, has resulted in a high level of diversity.

The basic characteristics of its cultivation systems are its year round cultivation and staggered

PEOPLE AND PLANTS: CULTIVATION

harvests. The greater share of the garden work is undertaken by women, and the propagation is carried out by cuttings taken from productive gardens. The compounds are often owned collectively and operated for a number of crop cycles by kinship groups. As the crop can be harvested throughout the year, the garden also serves as a storage house. The tubers are consumed as human food and as animal feed.

The importance of sweet potato in Indonesia is surprising in many respects given the fact that it is of American origin, and probably arrived in the Indonesian Archipelago only about 400 years ago. Its spread into the highlands of the island of New Guinea came even later. There it met with an indigenous agriculture relying mainly on bananas taro, sugarcane, and yams, with an agricultural history dating back at least 5,000 years. Despite of this, however, the sweet potato became extremely popular: the famous 'checkerboard' gardens formed by sophisticated drainage ditches and raised beds — spotted from the first aeroplanes to cross the Baliem Valley in the 1940s — were planted with sweet potatoes.

The widespread cultivation of the sweet potato is due to two factors. Compared with taro, previously the most important tuber crop, sweet potato is far more adaptable to the prevailing environment, especially to drought and low temperature. In addition, it is suitable as fodder for pigs, which play an important part in the lives of these people. The adoption of sweet potato cultivation, therefore, enabled highland societies to expand their settlement range further into altitudes above 2,000 metres and to raise more pigs. It also allowed people to rely more on horticulture and less on hunting. The expansion of sweet potato agriculture on steeper slopes and at higher altitudes continues to this day, although cropping and land-use patterns are slowly starting to shift towards commercial crops such as vegetables and coffee.

FOOD PRODUCTS DERIVED FROM CASSAVA

Tape (fermented block cassava): cassava roots are cut into five to ten centimetre lengths, washed and half cooked. They are then inoculated with *tape* inoculum, covered with banana leaves and incubated for two to three days. The different types of microorganism plays an important role in the texture and flavour of *tape*: these include yeasts (*Saccharomyces, Hanselula, Endomycopsis* and *Candida*), lactic acid bacteria (*Pediococcus*) and amyotytic moulds (*Mucor, Amylomyces* and *Rhyzopus*).

Brem: fermented block cassava whose initial stages of processing are similar to *tape*. After fermentation, the slurry is pressed and moulded into small pieces and then dried.

Cassava flour: cassava roots are peeled and washed, then chipped and dried to produce *gaplek*. The chips can be stored or milled into flour. The flour may be used to produce chocolate-based products, such as cakes, biscuits and bread.

Tiwul: is made from *gaplek* which has previously been pulverised and sieved. The meal is kneaded into paste along with a little water, mixed with coconut or cane sugar, and steamed. This gritty material is served as a substitute for rice.

Gatot: *gaplek* is cut into pieces, steamed and spread out on a bamboo mat. The pieces are kept wet for two to three days with continual sprinkling of water, after which they turn black. They can be served after steaming.

«*Cassava* (Manihot esculenta): *A perennial shrub about one to five metres tall.*

Tapioca processing in Bogor, West Java.

Slope cultivation of sweet potato at 2,100 metres in the Baliem Valley, Irian Jaya.

MARKET FORMATION AND AGRICULTURE OF ROOT CROPS: 1900–1990

	1900	1940	1970	1990
Cassava				
Area Planted (,000 Ha)	130	1,040	1,397	1,308
Yield (ton/ha)	6	8	7.5	12.1
Sweet potato				
Area Planted (,000 Ha)	80	210	356.5	209.8
Yield (Ton/Ha)	6	7	6.1	14.0

Source: Bottema, 1995.

CASSAVA AND SWEET POTATO PRODUCTION AREAS OF 1992

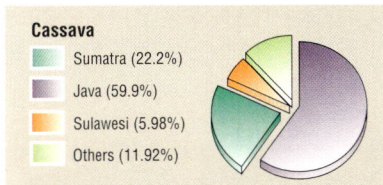

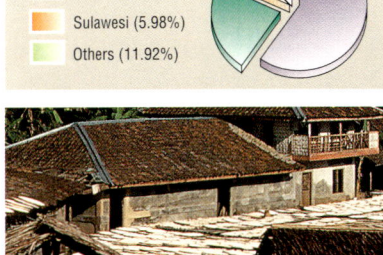

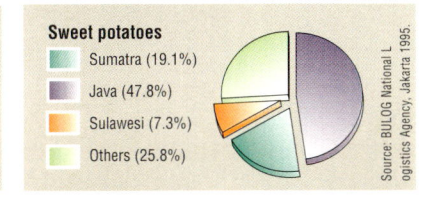

Cassava
- Sumatra (22.2%)
- Java (59.9%)
- Sulawesi (5.98%)
- Others (11.92%)

Sweet potatoes
- Sumatra (19.1%)
- Java (47.8%)
- Sulawesi (7.3%)
- Others (25.8%)

Source: BULOG National Logistics Agency, Jakarta 1995.

Fruits and Nuts of Indonesia

The islands of Indonesia have the greatest diversity of fruits in the world, with over 400 species of edible fruits, nuts and seeds. Not all products are found to the same extent throughout the Archipelago: some occur naturally, whereas others are cultivated. In addition to being eaten in their raw state, many fruits, particularly when young, are used in pickles, or dried and added to other dishes for extra flavour; the flowers and leaves of several fruit trees are also eaten.

FRUIT PRODUCTION FOR 1992

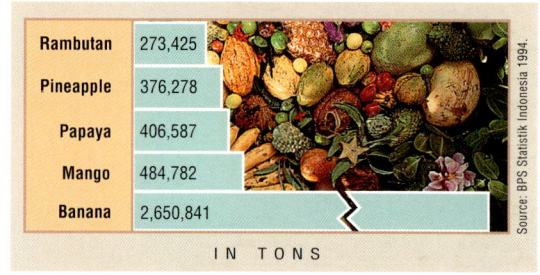

Rambutan	273,425
Pineapple	376,278
Papaya	406,587
Mango	484,782
Banana	2,650,841

IN TONS

Source: BPS Statistik Indonesia 1994.

Indonesian stamps featuring local fruits — gandaria (Bouea macrophylla) and ramboutan (Nephelium rambutan-ake).

Markisa, a popular drink prepared from juice concentrate extracted from the passion fruit (Passiflora edulis).

A Cornucopia of Fruits

The most well-known fruits are seen in markets throughout the year — banana, pineapple, watermelon and papaya, all of which were introduced many decades ago. The popularity of fruit can be seen from the numerous markets and stalls piled high with spiky football-sized durians, angled yellow starfruits, 'dragon's egg' salaks, hairy red rambutans, huge jackfruits with their redolent yellow flesh and the deliciously refreshing mangosteen.

While most travellers are familiar with these common cultivated fruits, found in Southeast Asia, few realise that a great range of lesser-known native fruits, many of which are brought to the markets only sporadically, exists. It may come as a surprise, for instance, that seven wild and edible species of durian are found in Indonesia.

One highly versatile fruit is the papaya. Young leaves and fruit are often eaten as a vegetable or pickled. In Sulawesi, the flowers of the male trees are used to make a delicious soup.

Mango

The mango is one of the world's oldest cultivated fruits, and most of those seen in the market are varieties of the common Indian mango (*Mangifera indica*), introduced to Indonesia centuries ago. In addition to this popular, widespread species, other wild species, play an important part in local cuisine.

In Java and Kalimantan, the closely related *binjai* (*M. caesia*) and *kemanga* (*M. kemanga*), with their delicate lilac-blue flowers are common trees. The large brownish-yellow fruit have a rather fibrous white flesh with a unique sweet-sour flavour when fresh. Young fruits are also used to make pickle and the young leaves of *kemanga* are eaten as a vegetable. In Kalimantan, the small, dark-purple kasturi mango (*M. casturi*), which has a juicy, sweet, orange flesh, is well-known around Banjarmasin, but remains almost unknown in other parts of Indonesia.

Durian

Durians are another group of plants with a wide variety of edible species. Only one, *Durio zibethinus*, the 'King of Fruits', is well-known. Famed for its unforgettable smell, it is either loved or hated on first acquaintance and nearly every early traveller to the East was impressed by this fruit — one way or the other. Eric Mjoberg wrote in 1930, 'One thing is certain... our first meeting with durian fruit is a memory which we carry with us to our life's end!'.

As with wild mangoes, the wild relatives of the durian have been largely ignored, yet several varieties possess just as much potential for recognition as tasty fruits in their own right. *Durio kutejensis*, a common wild species from Kalimantan, is prized by those who appreciate the thick, cheesy, rich orange flesh. Another wild species, *D. oxleyanus* from Borneo and Sumatra, has

A Fruiting branch of durian (*Durio zibethinus*), the flowering to the harvesting stage takes five to 5 1/2 months
B Jackfruit (*Artocarpus heterophyllus*) variety Dulang characterised by its thick flesh
C Snakefruit (*Salacca zalacca*) an endemic species which is in season all year round
D Duku (*Lansium domesticum*) variety Duku
E Mangosteen (*Garcinia mangostana*), known locally as 'an honest fruit'

PEOPLE AND PLANTS : CULTIVATION

❶ *Papaya* (Carica papaya)
❷ *Watermelon* (Citrullus lanatus)
❸ *Avocado* (Persea americana) ❹ *Pomelo* (Citrus maxima)
❺ *Starfruit* (Averrhoa carambola) ❻ *Durian* (Durio graveolens)
❼ *Calamondin* (Citrus mitis)
❽ *Litchi* (Litchi chinesis)
❾ *Guava* (Psidium guajava)
❿ *Waterapple* (Syzygium javanica)

«*Cashew nut and apple* (Anacardium occidentale).

Candle nut (Aleurites moluccana), *a native to Indonesia.*

a milder flavour and less vigorous aroma than *D. zibethinus* and is actually preferred by many who have tasted both. *D. graveolens*, also from Borneo and Sumatra, has smaller fruits with an orange to pink or even red flesh and with a delicate flavour like that of avocado but it has almost no smell.

People are by no means the only ones who enjoy gorging on durian fruit. Primates, wild pig, deer and even tigers are known to actively search for durian fruit in season. Elephants are said to swallow the spiny fruit whole, having first rolled them over and over in dead leaves to cover the spines.

Mangosteen

The mangosteen, *Garcinia mangostana*, with its unmistakable round, purple-skinned fruits, was called the 'Queen of Fruits' by early travellers to the East. Rumphuis in 1747 stated that 'It is remarkable that the mangostan (sic) is given with safety in almost every disorder.' He adds, 'When sick people have no relish for any other food they generally eat this with delight, but should they refuse it, their recovery is no longer expected.' The dried rinds of mangosteen were once an important part of the Chinese pharmacopoeia and are still used locally for dysentery today.

Rambutan and Langsat

The rambutan (*Nephelium lappaceum*) is probably one of the commonest fruits grown. Rambutan trees are a familiar sight in many villages. The fruit are easily recognisable by the red skins covered with long stiff hair-like spines. Several cultivars are now recognised — some with yellow or even green skins. Closely related to the rambutan is the pulasan (*Nephelium ramboutan-ake*) with stubby spines and a thicker skin; it is dark purple, yellow or green in colour. The pulasan is usually considered even more delicious than the rambutan by aficionados, though it is not so well-known.

Often growing with the rambutan, and usually fruiting at the same time, is the langsat (*Lansium domesticum*). The langsat is a small tree with long strings of the pale yellowish fruit dangling from the branches. Dried langsat skins are sometimes burnt as an insect repellent.

Salak

Another popular fruit in Indonesia, but one that is less well known outside the Archipelago is the *salak*, which is produced by the stemless palm *Salacca edulis*. These fruit with tough, scaly skin are often called 'dragon's eggs'. The white, crisp flesh has an almost nutty flavour. It is one of the few Indonesian fruits that is found in international hotels along with the popular banana and papaya.

Nuts and Seeds

Although the introduced cashew is cultivated in drier areas of Indonesia, two other trees are also commonly cultivated for their seeds. *Melinjo* (*Gnetum gnemon*) is a primitive tree, grown particularly in Java where the seeds are made into a tasty fried biscuit or *keropok*. *Petai* seeds from the leguminous tree *Parkia speciosa* are often cooked together with rice or added to sambal (a delicious chilli paste) and is especially liked by the Javanese.

The Javanese are also famed for their *tempe*, which is made from cultured soyabeans (in a way similar to making cheese). Many species of Artocarpus seeds are also roasted or boiled and consumed as a delicacy. The Java almond or kenari nuts (*Canarium indicum* and *C. vulgare*) and the wild meritus (*C. megalanthum* and *C. decumanum*) also produce tasty edible nuts, not unlike the more well-known pili nut of the Philippines.

Palms: an Introduction

Palms are regarded as being second only to the grasses in terms of their economic importance. Throughout the Indonesian Archipelago, both cultivated and wild species of palm are used extensively for a vast range of purposes including food, house construction, magic, medicine, weaving, clothing, decoration, and even writing materials — the earliest literature in the region was written on the leaves of the lontar palm (Borassus flabellifer).

Nibong, Oncosperma tigillarium, *also called the feathered palm, growing in the Sunda Straits.*

A hen nesting in a shelter woven from palm leaves in Sumba, East Nusa Tenggara.

Detailed patterns on the frond of the Livistona rotundifolia, *in North Sulawesi.*

An Abundance of Palms

Palms are a conspicuous element in almost all Indonesian landscapes, whether natural or manmade. Their ubiquity is reflected in the immense importance of palms and palm products in the Indonesian economy, both in terms of international trade and at a village level, especially in subsistence communities in rural areas. Important cultivated palm species include the oil palm (*Elaeis guineensis*), which was introduced from Africa and is now grown on large plantations, especially in Sumatra; the domesticated coconut (*Cocos nucifera*), a possible native of Indonesia along its northern Pacific fringe; and the betel nut palm (*Areca catechu*). Semi-domesticated species include the sugar palm (*Arenga pinnata*), sago palm (*Metroxylon sagu*) and lontar palm. Then there are the truly wild species, such as nipah (*Nypa fruticans*) and a vast number of climbing varieties which are collected for rattan.

Plants of Great Local Importance

In some areas, the local population is almost entirely dependent on palms for their daily existence. Communities in the eastern and drier parts of Nusa Tenggara rely heavily on the lontar palm. Villagers on the island of Roti, for example, tap its sweet sap and use it as a liquid staple. Elsewhere, coastal communities exploit coconut trees for building materials as well as for their edible products.

Before the advent of rice cultivation in island Southeast Asia, palm starch, or sago, obtained from the pith of the trunk of several species of palm, provided the staple diet for the peoples of the region. The situation remains unchanged in many areas today. The most important of the sago palms is *Metroxylon sagu*, which is native to Papuasia (including Irian Jaya), but now widespread throughout the wetter parts of Indonesia. It is cultivated both for its starch and for its leaves which provide the raw material for one of the most durable palm thatches. Other starch producing palms include the gebang (*Corypha utan*), the sugar palm and the wild Bornean sago palm (*Eugeissona utilis*), which provides the staple food of the nomadic Penan people of Borneo. In rice growing areas, these starch-producing palms provide an important carbohydrate supplement in periods of rice shortages.

Palm Diversity

Indonesia has one of the most diverse palm floras in the world with over 480 species belonging to 46 genera, almost one-fifth of all palm species and genera. Representatives of four out of the six evolutionary lines recognised within the family can be found occurring naturally within Indonesia. Native palms range from the majestic tree palms, which can contribute to the forest canopy, to squat palms of the forest undergrowth and climbers (rattans). Much of this diversity remains poorly known and requires additional research. The extraordinary variety of Indonesian palms is probably a reflection of the complex geological history of the area. In certain types of natural vegetation, for

PEOPLE AND PLANTS: CULTIVATION

example, humid lowland rain forest in Borneo, the palm flora may consist of 40 or 50 different species growing in the same area, of which at least half may be rattans. Even so, it is rare to find a situation where palm species completely dominate the vegetation, a notable exception being the nipah palm which occurs in vast, natural stands in coastal and estuarine habitats.

Natural History

Little is known of the natural history of native Indonesian palms. The few studies which have been conducted indicate a diverse and intricate range of pollination and dispersal strategies. Their abundance in natural vegetation, together with their visual distinctiveness as a family, makes them attractive subjects for future research.

Future Potential

While the native palms of Indonesia play a vital role in the rural economy of Indonesia few, if any, are exploited to their full commercial potential. In this respect, there is enormous scope for increasing the production of palm starch for industrial purposes and for expanding the production of palm sugar. The manufacture of fuel alcohol from palms such as nipa, offers an exciting possibility for the sustainable production of fuel from managed natural vegetation if petroleum becomes scarce in the future. Meanwhile, the rattan industry is also expanding as a result of demand from the furniture trade. Nor should the potential of the trade in ornamental palms be underestimated; at present there are few controls on the gathering of species in the wild, but there is an increasing demand for many of the beautiful native palms of Indonesia in international horticultural trade. This situation needs to be carefully monitored lest an uncontrolled trade threatens the future of some of the rarer species.

*The pandan, with its slender stems, is often mistaken for a palm.

*A fruit cluster of the lontar palm (*Borassus flabellifer*), which is produced in great quantities in East Java.*

*An elegant gebang (*Corypha utan*) in the Baluran Game Reserve of East Java.*

COMPOSITION OF SOME COMMON INDONESIAN PALMS

An array of palms found in diverse ecological niches ranging from brackish water to rain forests of Indonesia:
❶ *Sago palm* (Metroxylon sagu)
❷ *Nipa* (Nypa fruticans)
❸ *Sugar palm* (Arenga pinnata)
❹ *Oil palm* (Elaeis guineensis)
❺ *Rattan* (Calamus sp.)
❻ *Lontar palm* (Borassus flabellifer)
❼ *Sealing Wax palm* (Cyrtostachys renda)
❽ *Coconut* (Cocos nucifera)
❾ *Pandan** (Pandanus sp.)
❿ *Gebang* (Corypha utan)

Palms: Coconut and Areca Nut

Coconut and betelnut palms both play an important role in the lives of Indonesians. They are grown by most rural households and so are a common sight throughout the country. In contrast to the betelnut, which is only cultivated on a small scale, the coconut is grown both on a local and commercial scale. In addition to their fruit, a range of other products are derived from these trees.

»»*Workers on a coconut plantation.*

»»*Coconut husks are used as fuel for a limestone kiln in Bali.*

Coconut

The coconut (*Cocos nucifera*) is probably the most economically important and widely distributed of all palms. This wonderfully useful plant probably originated in eastern Indonesia or Melanesia (although many believe it is of heavenly origin), but is now found in nearly all tropical countries. The fruits disperse naturally by floating in ocean currents, but the present distribution is largely a result of deliberate planting by people. It is a large palm, growing up to 30 metres, with a slender stem which is slightly swollen at the base, often curved or slanted with a crown of spirally arranged, pinnate leaves. The inflorescence is produced in the leaf axils and bears separate male and female flowers. The first flowers and fruit are produced after six years and fruiting continues up to 70 years. Fruits mature in about 12 months, by which time they reach an average weight of 1.5 kilogrammes, ripening to green, brown or yellow–orange.

Wide Range of Uses

Almost all parts of the coconut are used, but it is the fruits which are the most important commercially. The main product is oil, which is extracted from the flesh and used mainly for cooking and in the manufacturing of soap. Oil can be produced domestically or the dried flesh, which is known as copra, sold to oil mills. Much of the oil pressing occurs within Indonesia, but copra is also exported. The flesh can also be eaten fresh, either when it is mature or when young and still jelly–like. The mature flesh is grated to extract coconut milk, that vital ingredient of so many Indonesian culinary dishes. The nut cavity, is partially filled with a liquid — coconut water — which makes a refreshing drink. Coconut milk has the same

Trained pig–tailed macaques (Macaca nemestrina) picking coconuts.

COCONUT PALMS (*Cocos nucifera*):
- **A** Late fruiting (six to eight years) tall variety, 'Kelap Dalam'
- **B** Flower stalk of the coconut palm
- **C** Halved young and matured coconut
- **D** Early fruiting (three to four years) dwarf variety, 'Kelapa Genjah'

PESTS
1. The rhinoceros beetle bores into coconut trees through the leaf bases to feed on exuded sap.
2. The plantain squirrel chews through coconuts causing considerable loss to farmers.

COCONUT PRODUCTS:

LEAVES — woven into hats, fans and baskets and also offerings and temple decorations

HUSK — the main source of coir, used in making mats and ropes

COCONUT MEAT — the flesh is used in ice cream, cosmetics and oil, and the milk and grate are included in local dishes.

HARD INNER LAYER — used to make charcoal filters and trinkets such as buttons.

proportion of salts as human blood plasma and so can be used as a sterile source of surgical drip in an emergency! The shell of the nut has a range of uses, for example, making household pots and utensils, or as a source of charcoal or powder for industrial products. Coir, the fibre on the husks, is used to make ropes, mats and brushes. Thus the fruit alone provides food, drink and fibre. Additional products are derived from other parts of the coconut. The flower stem can be tapped for the sweet sap, drunk either fresh or fermented, or concentrated to produce palm sugar. The trunks are used as fuel and also for timber, although the wood is difficult to work because it is very hard. The leaves are commonly used for thatch and the leaflets made into weaving mats, baskets and hats.

Monkeys as Harvesters

Indonesia is one of the largest producers of coconut in the world, accounting for almost one-third of total global production. This supplies the enormous home market, as well as products for export. A great deal of research has been carried out on the coconut, aimed at improving yields and productivity. Modern dwarf cultivars start flowering within three years, unlike tall varieties which take about six years to flower. More recently, strains have been developed by crossing the tall and short varieties to produce an early flowerer which will also crop as heavily as the tall varieties. Coconuts may be harvested in a variety of ways. In some parts of Indonesia macaques (*Macaca nemestrina*) are trained to pick coconuts either by biting through the stalk or twisting the fruit until it falls. In Sumatra, alone, about 80 per cent of the coconut gardens are harvested by trained monkeys.

Areca Nut

The betelnut or areca nut (*Areca catechu*), known locally as *pinang sirih,* occurs throughout Southeast Asia, growing best in humid lowlands. In Indonesia, this palm is found in home gardens, where one or two trees are cultivated to provide for household purposes only. It is a slender palm with featherlike leaves, growing to a height of 12 metres. The stem is grey and ringed with leaf scars and bears a well-defined crownshaft — the conspicuous 'cylinder' formed at the top of the stem by the leaf bases. The fruits, for which this palm is most highly valued are egg-shaped and ripen to a bright orange colour.

Stimulating Wad

The fruits of this palm are harvested for their seed which contain the alkaloid arecaine, which is a mild stimulant. It is this property which has led to the widespread practice of betel-chewing. The seed is sliced and wrapped in the leaves of the betel pepper (*Piper betle*), together with lime to form a chewing wad which, when held in the mouth, is said to relieve hunger and tiredness. Betel chewing has a long history in the region, and has been recorded since the 13th century. As well as being an everyday practice, it is also of ceremonial significance. Many artefacts are produced which are associated with betel chewing, such as boxes and scissors. Betelnuts are still widely sold in the markets but chewing betel is becoming less popular, particularly among the younger generation. In part this is because of the staining properties of the betelnut, which turns saliva red and stains a person's teeth black.

Batik Dye

The staining property of betelnuts is exploited in the batik industry, which uses the seeds as a source of red dye. The seed is also used medicinally, for the treatment of leprosy, anaemia, obesity, for expelling worms and as a tooth powder. Other parts of this palm can be used, although these are of secondary importance; for example thatch is made from the leaves and the trunk is used as a source of wood.

A horse-shaped betel cutter, a popular form found in Sumatra, Madura and Bali. Betel cutters are used to slice and pare shavings from the hard areca nut. The basic cutter consist of two arms pivoted on a simple rivet or hinge.

THE ARECA NUT (*Areca catechu*)
1. Betel palm — tall slender feather palm with distinct crown consisting of arched leaves and inflorescence below the crown shafts.
2. Fruiting branch with ovoid fruits, narrow at the apex.
3. Section of young and mature fruit showing the ruminate endosperm.
4. Betel set containing
   - A Betel leaves
   - B Gambier
   - C Areca nut
   - D Lime
   - E Mortar for crushing the various components into a mixture for chewing
   - F Tobacco

A man with his saliva stained red and blackened teeth, exhibiting the long term consequences of betel chewing.

Palms: Sugar, Lontar and Nipa

The sugar, lontar and nipa palms are all excellent examples of plants with a multitude of uses. Nearly every part of these plants — fruit, leaves, trunks, roots and sap — is useful to people. For all three species, it is the sweet sap which has the greatest economic value; there is considerable potential for further exploitation of this resource for the production of alcohol and sugar.

»»*Sugar palms (*Arenga pinnata*) in Kalimantan in flower.*

»»*The fan-shaped Lontar (*Borassus flabellifer*) in Bali.*

A lontar palm as depicted on the reliefs at Borobudur.

»»*Old manuscripts written on lontar leaf.*

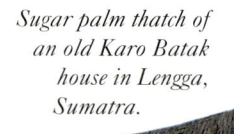

Sugar palm thatch of an old Karo Batak house in Lengga, Sumatra.

Sugar Palm

Sugar palm (*Arenga pinnata*) is thought to be indigenous to the Indo–Malayan Archipelago. This striking palm is frequently found growing near villages where it is informally cultivated. The solitary trunks, which are covered with persistent leaf bases and long black fibres, bear a dense crown of pinnate leaves. After a period of vegetative growth, pendulous inflorescences are produced from the top of the trunk downwards. These are usually unisexual, the female flowers being produced first. The palm dies after flowering and fruiting.

The most important product of this palm is its sugary sap — hence its common name. The sap exudes from the cut stalk of the inflorescence; an average yield of about six litres can be obtained each day. This sap makes a refreshing drink *neera*, but may also be boiled down to make *gula merah*, a dark red sugar. Palm wine, or *tuak*, is made by allowing fermentation of the sap, and vinegar is produced if this process is continued further. Distillation of *tuak* produces the potent drink, *arak*, and also alcohol for industrial purposes. The sugar palm is also a valuable source of *ijuk*, the fibre which covers the trunk of this palm. It is used throughout Indonesia because of its great durability, making excellent cordage, thatching, matting and brushes. Fibre is also recovered from the roots, trunk pith and leaf stalks, although this is of a lower quality.

In addition to the above, the sugar palm provides a range of foodstuffs. The shoot apex is a source of palm cabbage and good quality sago can be extracted from the stem. Immature fruits are collected for the edible, gelatinous endosperm. The latter becomes very hard with maturity, and is then used for carving. The fruits are also used as a fish poison, since the skins are toxic.

A popular delicacy are the grubs of the palm beetle, which are reared on fallen stems. Recent interest in this species has led to the development of commercial plantations.

This highly productive and versatile palm could play an increasingly important role in the economy of Indonesia in future years.

Lontar Palm

The *lontar* or palmyra palm (*Borassus flabellifer*) is found throughout Southeast Asia but is confined to dry forest and savanna. This large palm produces fan-shaped leaves and a dense fringe of adventitious roots at the base of the trunk. This species has separate male and female plants, and it is the fruit of the female palms which are most intensively used. Most of the fruits are harvested when they are immature. At this stage, the endosperm is jelly-like and is popular as a sweetmeat, eaten fresh or cooked with sugar, or mixed with coconut milk. In some areas, 50 per cent of the fruits may be harvested for their endosperm. The ripe fruits are gathered for the sweet, orange pulp extracted from the mesocarp. A mature tree may produce 300 fruits annually, so there is some potential for sugar and carbohydrate extraction from excess harvests. Tapping the inflorescence for sap, which is also widespread, may, however, produce a better income for farmers. At present, the sap is used to produce sugar, beverages and vinegar for local markets.

On a smaller scale, the palm heart is eaten, as are the tuberous seedlings which are rich in starch. In the past, the leaves were greatly valued as a writing material. Today they are used for basketry and thatch, and also provide a valuable fibre which is extracted from the petioles and leaf bases. Despite the multiple uses of lontar, it remains under-exploited. Efforts are underway to improve the harvesting and manufacture of its products, which will enable the development of this resource.

Nipa Palm

The nipa palm *(Nypa fruticans)* has an ancient fossil history which dates back to the Upper Cretaceous. *Nypa* pollen is one of the earliest to be identified as belonging to a flowering plant. At one time this species was widely distributed throughout the tropics, but the current distribution of this palm is greatly contracted. Today the nipa palm is an important component of the mangrove forests of Southeast Asia and the west Pacific. It grows along river banks and in shallow, brackish lagoons often forming vast, almost pure, stands.

Nipa is the only member of its sub-family, the Nypoideae, reflecting its distinctiveness within the palm family. Unlike other palms, the stem of nipa is prostrate, usually lying submerged in the mud. The stem branches dichotomously, each terminal shoot producing a cluster of pinnate leaves. This creeping habit serves to stabilise estuarine mud and thus prevents coastal erosion. The inflorescence is also unique in the palm family in that it is held erect, and bears a terminal spike of female flowers with lateral branches bearing the male flowers. The fruits are tightly clustered in a large round head which floats on the water surface. When ripe, the fruits begin to germinate, causing the head to break up. The individual fruits are then dispersed by the tide.

The nipa palm plays a vital role in the lives of people living near mangrove forests. Historically this palm has been the source of a wide range of products, some of which today form the basis of cottage industries and commercial operations. Its primary use is for thatch and there is a considerable trade in nipa leaves which are prized above all others for this purpose. The leaves are first dried and the midrib removed. They are then folded over a bamboo or wooden rod and sewn in lengths with vine to make a 'shingle', either for roofing or walls. The leaves are also woven into hats, baskets, mats and bags.

Other products include fishing floats made from the petioles, rope from the twisted midribs, and cigarette papers from young leaflets.

The endosperm of the fruits are harvested for food when they are young and gelatinous while the apical buds are eaten as a vegetable. The nipa palm is also tapped for its sweet sap, but is greatly under-exploited as a source of sugar. The yield of sap is much lower than that of the sugar palm, with an average tree producing 1.25 litres per day. Nevertheless, Indonesia has vast areas of swamp, and so there is great potential for developing this resource. Unlike sugarcane, the tapping of nipa provides continuous labour throughout the year and does not compete with other crops for agricultural land.

The process of making tuak syrup, using buckets made from palm leaves of the versatile lontar.

Nypa fruticans *growing in its habitat in the brackish waters of Indonesia:*
❶ *Compound leaf with regularly repeated segments*
❷ *Inflorescence showing both the terminal female spike and lateral branches of male flower*
❸ *to* ❺ *Fruiting heads in various stages of development. Mature heads eventually detach and are dispersed by water*

«« *A lontar tapper crushing the tip of a bud to obtain a flow of sap.*

Palms: Sago Cultivation

The sago palm (Metroxylon sagu) is one of the oldest food crops in Southeast Asia and Micronesia. It is probably native to Maluku and the island of New Guinea but its range now stretches from the Mentawai Islands to southern Thailand, southern Philippines, and the islands of the western Pacific. Sago palms possess a competitive advantage over other trees in swampy areas. The island of New Guinea is now considered to be the centre of diversity of these palms.

A clump of sago palms at different stages of rosette–growth: (anti–clockwise)
❶ *and* ❷ *Various stages of trunk formation.*
❸ *Full trunk growth is achieved after 4 1/2 years and the height reached is about 10 metres.*
❹ *The inflorescence stage is reached in another 3 years, during which the palm bursts into a dramatic bloom.*

Asmat men transporting sago pith obtained from felled palms in make–shift carriers made from palm leaves.

Collecting sago worms, larvae of the weevil Rhynchoporus ferrugineus, *a prized delicacy from rotting old trunks.*

Life Cycle

The sago palm grows in clumps. When the seed germinates, a rosette of leaves is formed without a trunk. Leaves grow from the centre, while those on the outside die. They slowly increase in number as well as in size and, under optimum conditions, after three and a half years the growing point is sufficiently large for trunk formation to begin. Starch is stored in the trunk for later use in forming the huge, flowering structure. This inflorescence heralds the death of the palm. Both the size of the inflorescence and the number of seeds produced depend on the amount of starch stored in the trunk.

Harvesting Sago

Indigenous people derive their carbohydrate from the trunk of the palm, using the starch from the pith as a staple food. In extremely wet conditions, wild and uncultivated sago palms usually produce about 100 kilogrammes of dry starch. Through a selection process, however, some trees are capable of producing up to 600 kilogrammes of dry starch from a single trunk. These trees are cultivated on drier ground and other trees are felled so that they do not shade the young palms. The original rosette and, later, the trunk form suckers which have exactly the same genetic make–up as the parent plant. Large suckers of these plants are chosen as planting materials.

Harvesting sago is a relatively simple process; the palm is felled and the bark removed to reveal the soft, central pith, which is pulverised. The pith is washed in a large sieve to remove the starch which is collected as it settles.

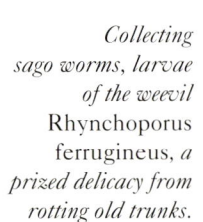

A sago tree ceremony performed by the Asmat in Irian Jaya before the felling of the palms.

Clockwise: Washing and preparing sago in the island of New Guinea. Sago flour is transferred to containers made from stitched sago leaflets. Containers of sago being pushed into the mud on the Mentawei islands. In anaerobic conditions the sflour will decompose only slowly.

The starch is usually white, or slightly discoloured, and is saved in packages made of sago palm leaves. All other materials are left in the field and thus returned to the soil. This whole process takes a family of four only two or three days, excluding the time taken to reach the site, and results in 200 kilogrammes of starch from a fairly good trunk. This provides sufficient carbohydrate for a family of four persons for about a 100 days.

If the starch is saved without further drying, it quickly spoils and takes on a rather unappetising lactic acid smell. This may be avoided by storing it under water where the lack of oxygen prevents quick spoilage. Local people prepare sago in a variety of ways: stirred with hot water and eaten as a sticky porridge, resembling glue; or baked in sago leaves to give thin sticks which taste like fresh bread when warm but are less appetising if eaten later. Sago starch can also be heated in banana leaves, mixed with other kinds of food, for example, fish or ground kenari (*Canarium commune*) seeds.

If the trunk is harvested before the flower develops, the large, soft, sweet growing point can be eaten, as with many palms. Grubs or sago worms may develop in old, rotting trunks and in remnants of starch processing. These are the larvae of the weevil *Rhynchophorus ferrugineus* which are a prized food for sago growers, containing some 5 per cent protein and 20 per cent fat. The trunks of mature sago palms are sometimes split and wedged open to enable the weevils to lay their eggs.

Mature sago leaves are used for thatch. The leaflets are removed, doubled around a stick and stitched. Sago makes the most durable of any palm thatch. The leaf stems are used as building material and the bark is often used as flooring.

Commercial Growth

In the last 20 years, interest has developed in using the sago palm as a commercial crop. As starch finds a ready market, it can be used in all kinds of food: noodles, biscuits, even in bread as a partial substitute for imported wheat flour. Industrially, it is used in the paper and clothing industry, in medicines, in the preparation of glues and high-fructose syrup. In Java, most sago starch is used in Cirebon in the manufacture of noodles. Sago starch could also be used for the production of fuel alcohol. At present, however, the price of fossil fuel is too low to make the production of fuel alcohol a viable proposition, while starch is more lucrative as it can fetch higher prices in other markets.

Triggered by economic possibility, attempts are being made to start large-scale sago plantations on peat soils in Sarawak and to exploit natural strands of sago palm as a business venture on Halmahera and the Bird's Head Peninsula in Irian Jaya. In these operations, the trunks are being cut in the field and floated out to a factory for processing. Quite a number of difficulties have to be overcome in this process, including the problem of rapid nutrient depletion in the field caused by the removal of the trunks to a central factory, and finding the best way of transporting the trunks.

In the past the sago palm was considered to be a 'poor man's crop' as it was only used as a food crop in the most remote parts of Indonesia. As a result, little attention was given to this most useful plant before the mid-1970s.

A cartoon from a Dutch newspaper, De Volkskrant. 'I never leave without some reserve fuel'.

An old picture of a sago manufactory where sago starch was extracted by primitive means. First the log was cut lengthwise and the inner tissue rasped into a coarse 'sawdust-like' meal. Then the starch was washed out of the meal and purified by repeated washing. The resulting flour was formed into pellets or grains.

Gingers

Gingers are aromatic herbs found throughout the world's wet, tropical regions. A total of 46 genera has been described worldwide, of which 21 are found in Indonesia. Among them are sources of many of the well-known spices, medicines and dyes, as well as beautiful and unusual flowering ornamental plants, some having the added attraction of an exotic perfume.

Shapes and Sizes

A member of the ginger family is generally quite easy to recognise, but precise identification of the genus or species is often much more difficult. The leaves of most gingers emerge alternately from the stem, but those of *Costus* (sometimes placed in a separate family) has leaves arranged spirally on the stem. Ginger flowers are orchid-like and short-lived, and are unique in having just one functional stamen; and the accompanying staminodes and usually-lobed labellum (lip) being more like petals. The inflorescences, in which the flowers are borne, have a wide range of forms and can be terminal on leafy shoots or basal on leaf-less lateral shoots. Some, such as species of *Alpinia* and *Etlingera*, gain a height of six metres or more, while diminutive species of *Boesenbergia* and *Kaempferia* of the lowland forest floor are just a few centimetres tall.

Ornamental Gingers

The flowers of most gingers are rather small and insignificant but they often have colourful bracts (modified leaves surrounding the flower), and it is for these that many indigenous Indonesian gingers are grown in gardens. Many species grow well in pots or borders in shady conditions, although the tall species are not appropriate for most gardens. Among those species grown in gardens are *Curcuma petiolata*, *C. purpurascens*, *C. zedoaria* and *C. xanthorhiza*, all of which are relatives of the turmeric and have colourful bracts. Some species are out in all seasons, such as *Costus speciosus*, which has large white flowers and bright red cone-shaped bracts but no fragrance. It is interesting because it contains the steroid diosgenin from which synthetic sex hormones can be made.

Selamat Makan

A great variety of spices are used in Indonesian cooking, and those from the ginger family are among the most important for flavouring meat and vegetables, making the food such a gastronomic delight. By far the most important species is turmeric or *kunyit* (*Curcuma longa*), the origin of which is unknown. Its most famous use is in making the mountain of 'yellow rice' used as the centrepiece of a Javanese *selamatan*, or celebration for a new child, as well as religious festivals and numerous other domestic events. Unlike many spices, it can be used either fresh, dried or as a ground powder. Turmeric is also used as a dye for cotton,

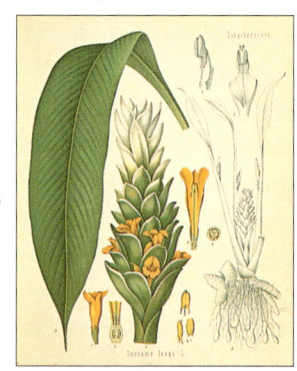

The habit of ginger flowers can differ even within a genus (from top): Costus speciosus *(top) has terminal flowers whereas* Costus globosus *(bottom) has basal flowers.*

Bags of instant ginger tea, a traditional health drink which warms the body and eliminates body aches and fatigue.

»»*An important spice in Indonesia, turmeric (*Curcuma longa*) is characterised by its large aromatic deep yellow rhizome*

IDENTIFICATION PROBLEMS

In the wild it is almost impossible, even for an expert, to ascertain to which genus a non-flowering ginger plant belongs because the leafy shoots can be very similar, even in distantly-related genera. One can also be fooled into thinking that a particular clump of ginger plants is not flowering, as the flowers may appear metres away from the clump, arising from the underground fleshy rhizomes or from long roots. This is found in *Hornstedtia* (top) and *Etlingera* (bottom). In other genera, such as *Elettaria*, the flowers are borne on elongated flowering shoots which bury themselves in the leaf litter and again appear at a distance from the main plant.

The flowers of *Hornstedtia scyphifera* (top) and *Etlingera littoralis* (bottom) emerge some distance away from the parent plant.

PEOPLE AND PLANTS: CULTIVATION

Riedelia *is a genus of around 60 species almost exclusively confined to Irian Jaya.* R. corallina *(left) together with* R. lanata *are grown in cultivation.*

»»*The main commercial ginger (*Zingiber officinale*) — a perennial herb widely cultivated for its rhizome.*

silk, as well as in margarines.
Other important species include *lengkuas* or *laos (Alpinia galanga)* which is used for flavouring food and as a medicine. It has an aromatic rhizome rather more medicinal in flavour than ginger, and is said to have actual medicinal properties, not just for humans but also for elephants which have lost their appetite. *Kencur (Kaempferia galanga)* has an aromatic rhizome which, when cut and dried, has a very strong flavour and must be used sparingly. It is said to originate from India, but is now widely cultivated in Southeast Asia, being used both for food and medicinal purposes.

Ginger was being used in Chinese cooking by 600 BC. *Jahe,* or the main commercial ginger *(Zingiber officinale),* has been widely cultivated for centuries. Although, its origin is unknown, the name being derived from the Sanskrit 'singa–bera' meaning 'horn-shaped' suggests it could have come from India. Ginger was thought to have been originally traded by the Arabs, who brought the spice from India before the Europeans traded in that area. Later, when the Portuguese took it to West Africa, and the Spaniards to Mexico, it subsequently spread across the American continent.

*Torch ginger (*Etlingera elatior*) is common in moist secondary growth. It produces flowering structures at the end of an erect, metre–tall stem.*

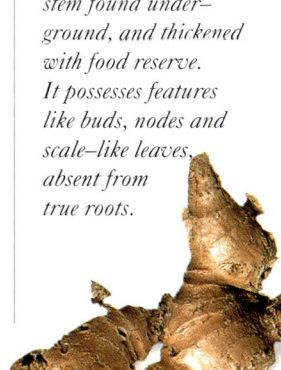

Ginger rhizome, an elongated horizontal stem found under-ground, and thickened with food reserve. It possesses features like buds, nodes and scale–like leaves, absent from true roots.

GINGER LILIES
The ginger lily, *Hedychium coronarium* (right), is probably the most commonly cultivated ginger in Indonesia, having a large terminal inflorescence of white flowers and a heady perfume. Two varieties of this species have been introduced into Java — these being varieties *flavescens* and *chrysoleucum* — both with yellow flowers. These species are commonly called garland flowers, ginger lilies or butterfly lilies. The flowers, although short–lived, are worn ornamentally and because of their beautiful scent. The white flowers of the common ginger lily and a relative, *H. horsfieldii* have a small lip and orange–red anthers. It is cultivated on account of its attractive bright red seeds surrounded by the yellow fruiting parts as demonstrated far right.

Nutmegs and Cloves

Nutmeg and clove, the most famous of Indonesian spices, used to be worth their weight in gold, and their sources were veiled in secrecy. Such was their importance in international trade that political intrigues, massacres, sieges, and invasions were all part of their early history in development and trade.

A clove plantation with the distinct landscape of slopes studded with cone–shaped clove trees, in Central Sulawesi.

The harvesting of the aromatically pungent flower buds of the red–flowered tropical evergreeen tree, Syzgium aromaticum.

Nutmeg (Myristica fragrans) *originates from the Banda Islands in Maluku:*
❶ *Halved fruit.*
❷ *Flowering and fruiting branch.*

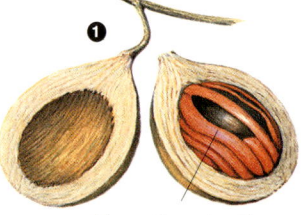

Mace, the red aril.

The Spice Trade

Spices, fine cloths, precious metals, gemstones and beautiful timbers have been valued by people throughout the ages. Spices were given to kings and emperors as gifts and tributes, and, with herbs, have formed the basis of pharmacopaeia and magic spells. Before the Portuguese reached the tiny Banda Islands in AD 1512, the spice trade was conducted by Arab, Javanese and Sumatran seafarers with links to Venetian middlemen. It was the hope of finding new spices, new supplies, better routes, fame and material wealth that gave rise to the great voyages of discovery undertaken by Europeans in the 15th and 16th centuries. Over the next three centuries the Portuguese, Spanish, Dutch and British fought to control or destroy the spice monopolies.

Flavoursome Nutmeg

Nutmeg (*Myristica fragrans*) is a small forest tree reaching only about 18 metres in height. It grows best where it is shaded by other taller trees. It has dark–grey bark, long glossy leaves, like a rhodo–dendron, small yellow flowers, and produces a fruit which looks somewhat like an apricot in size and colour. This splits open when ripe to reveal a shiny brown nut covered with a bright red net–like aril. This attracts flocks of the large imperial pigeon (*Ducula concinna*) which is the main species responsible for the dispersal of these seeds.

When dried, the aril is sold as mace, and the seed inside the nut is ground to make nutmeg powder. Nutmeg and mace are used as flavourings, preservatives, and in medicines, perfumery and cosmetics. Mace tastes similar to nutmeg but is more refined and therefore much more expensive.

The bark of nutmeg trees is characteristically thin and when cut or slashed oozes a watery pink sap which, on drying, darkens to the colour of dried blood. This accounts for the term *pendarah,* which in Indonesia is also applied to a *kris* that is said to have magical powers.

The Spice Wars

Precisely when the international trade in nutmeg began is not known, but the spice is recorded as having been imported to Constantinople in the 6th century, and it was widely known in Europe by the 12th century. Until the 18th century the nine Banda Islands, covering only 50 square kilometres, were the world's only source of nutmeg and mace. The source of these spices was a closely guarded secret of just a few Arab traders until the Portuguese found the islands in the early 16th century. At that time, the islands were an important entrepôt, trading also in cloves from Ternate and Tidore, bird of paradise feathers from the east, and slaves.

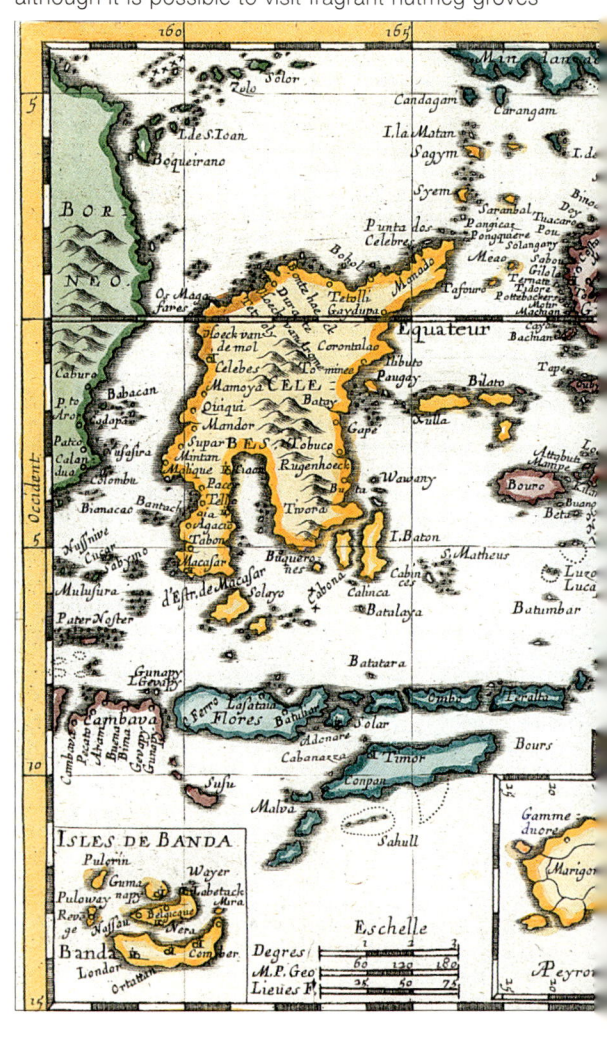

Later the Dutch and British came to build forts and many lives were lost in the struggle for power and profits. In order to monopolise the nutmeg trade, the Dutch felled all nutmeg trees that were not under their direct control, and murdered most of the male Banda population, dividing the land up among a motley collection of criminals and adventurers who used slave labour to tend and harvest the trees. Nearly 50 years after the massacre, the British ceded their former Banda island of Run in the west in exchange for the Dutch island of Manhattan, now the hub of New York City. Meanwhile the British took some nutmeg seedlings to Penang and then to the West Indies. The Banda Islands produce very little of the nutmeg traded today, with most coming from Sulawesi, Grenada and Sri Lanka, although it is possible to visit fragrant nutmeg groves under a canopy of *Canarium* trees on the largest Banda island of Lontar.

Tiny Cloves

The clove (*Syzygium aromaticum*) plant is a small understorey tree which reaches a height of 9 to 12 metres. Many parts of Indonesia have hillsides covered by these distinctive cone-shaped trees with their lime-green leaves suffused with red. Like most understorey trees, the clove is unable to regenerate in full sun and has a seed which is viable for only a short period. The trees are thus propagated by seeds in shaded nurseries and the seedlings planted out at two years. The cloves we use are actually the flowers, bunches of which grow at the end of the branches. The unopened flowers are carefully picked and then dried in the sun before being ground or used whole. An average eight metre tree will produce two to three kilogrammes of dried cloves each year.

Cloves were known to the Chinese of the 3rd century BC, who gave them a colourful name, 'odiferous nails'. Cloves were being used in Europe by the 4th century AD. Before the Dutch became directly involved, all the world's cloves came from just five small islands off the west coast of Halmahera: Ternate, Tidore, Moti, Makian and Bacan. Although nutmeg and mace were greaty sought after, it was cloves which dominated the trade in terms of monetary value. Today, North Sulawesi grows over one-quarter of the nation's cloves, followed by Central Java, Lampung and West Sumatra. Many parts of Indonesia have caught 'clove fever' with arable land and forest fringes being planted with the tree in the hope of reaping profits. The hopes of some farmers have been dashed, however, by low prices, pests such as boring beetles, and the serious 'clove leaf drop' and 'Sumatra' diseases which have destroyed thousands of hectares of plantations.

Going up in Smoke

While some cloves are used in cooking and dental care, the vast majority of these dried flowers are used by the *kretek* cigarette industry. Some 85 per cent of the nearly 30,000 tonnes of cloves Indonesia consumes each year are made into *kretek*. About one-quarter of this total is imported, primarily from Zanzibar. *Kretek* cigarettes were first smoked in Maluku in the 17th century, but modern production can be traced to the last century when Javanese smokers experimented with adding clove fragments to their cigarettes. This habit became so popular that today over 70,000 people are employed in the *kretek* cigarette industry which is responsible for using 50 per cent of the world's total clove production.

*A flowering branch of the clove (*Syzygium aromaticum*) which originates from the islands off the west coast of Halmahera in Maluku.*

«««Les Isles Molvcoves — depicting the Pacific islands where cloves grew in the 17th century.

Kretek cigarettes, are made from dried clove flowers. The main bulk of the cloves that are harvested form the basis of the cigarette industry.

The drying of the harvested clove buds in a village in South Sulawesi.

Pepper

For several hundred years, pepper was one of the world's most important trade commodities, and over half of it was grown in Indonesia. Even today, pepper still accounts for more than one-quarter of the entire world trade in spices.

Pepper (Piper nigrum) vines on Banka Island, Sumatra.

THE PRODUCTION OF BLACK AND WHITE PEPPER.
Black and white pepper are both obtained from the same vine, the difference is the time of harvest. Black pepper is produced by picking the fruits while they are still green. On the other hand, white pepper is produced by picking ripened fruits. Black pepper is the more pungent.

Cultivation of Peppers

Pepper is the dried fruit of an Indian vine (*Piper nigrum*), which is native to the forested slopes of the Western Ghats of southern India. This plant likes a hot, wet climate and is usually grown in lowland areas. Since it cannot tolerate waterlogging, it is cultivated on gentle slopes. If allowed to grow unhindered it can reach about ten metres in length, but most plants in cultivation are trained onto two metre posts.

Commercial pepper is grown from cuttings, because plants grown from seed take about seven years to bear useful quantities of fruit. Vines grown from cuttings are not allowed to produce flowering spikes until they are two years old, after which it takes about six months to produce ripe fruits. A vine typically has a productive life of 15 to 20 years.

Two forms of pepper are traded: black and white. Black pepper is produced by picking the fruits while still green, and piling them up, which allows fermentation to take place and increases their pungency. They are then dried in the sun. White pepper is produced by allowing the berries to ripen red before picking. They are then lightly crushed and soaked in water, and the skin and pulp rubbed off, before being dried in the sun. Each plant is picked over once a week during a two-month harvesting period. Towards the end of the period all the fruit is picked, irrespective of their state of maturity, in order that the next season's growth will appear at about the same time.

Pepper is cultivated in various ways, but the most intensive methods are found on Bangka Island off the southeast coast of Sumatra. Here the vines are grown unshaded on posts and carefully pruned to develop extensive lateral fruiting branches. Close attention is paid to weeding and fertilising with the result that 20 to 40 kilogrammes of pepper can be produced annually per plant. Serious soil erosion may occur, however, unless great care is taken with this process. Elsewhere, pepper is generally found growing beneath, and over, shade trees such as coffee. Under such conditions, the yields are somewhat less.

Use of Pepper

Until about 100 years ago the main uses of pepper were to flavour insipid food, to overcome the smell and taste of bad food, and to preserve meat. It is now used almost exclusively as a food flavouring: the alkaloid it contains, piperine, stimulates gastric juices and saliva. Its pungency is caused by a resin called chavicine which, being concentrated in the fruit wall, is more abundant in the black than the white form.

Pepper Through the Ages

Pepper is mentioned as a medicinal plant in Indian manuscripts dating back 3,000 years, and it was one of the earliest commodities traded between the East and Europe. Arabs brought it to the Mediterranean from India, where it was used by the ancient Greeks and Romans, both for its flavour and as a medicine to reduce fevers. It was probably introduced to Java from India between 100 BC and 600 AD. By the 15th century, it had become one of the world's most important commodities and it remained so until the mid-19th century. Its value was so high that it was commonly used for payment of taxes, rents and dowries. By the end of the 15th century, Sumatra

| Harvest | Fermentation | Dried |
| Harvest | Crushed | Soak in water | Remove skin and pulp | Dried |

PEOPLE AND PLANTS: CULTIVATION

Pedie on the northern coast of Aceh, Sumatra was one of the major centres for pepper exports during the 16th and 17th centuries.

was the main exporter of pepper to China. By the end of the following century, half of Europe's supply came from Aceh in north Sumatra, which, a century later was supplying pepper to half of the world.

Other major areas of pepper cultivation in the 17th century included west Sumatra, Lampung and west Java, with the Dutch and British establishing plantations in Java and Sumatra. The Dutch tried to gain a monopoly of the trade, as they had done with nutmeg and clove, but they never achieved it.

From the 16th to early 18th centuries the port of Banten, which lies some 75 kilometres west of present-day Jakarta, was a very successful trade centre for pepper and other commodities. The well-drained, sloping soil of the surrounding region was suited to its cultivation. The rewards of this trade were so great that local farmers were forced to grow a certain number of pepper vines per family, which they then had to sell exclusively to the Sultan's agents. This edict extended across the Sunda Straits to Lampung in southern Sumatra, which also came under the rule of the Sultan of Banten. The most intensive pepper growing areas were around Banten, where, eventually, serious soil erosion reduced the productivity of these fields and increased rates of sedimentation in the rivers prevented access to the port. As the quality of the environment deteriorated, so did Banten's role as an international centre of commerce. This process was hastened by the Dutch who had been trying to divert trade to their own centre at Batavia since the late 17th century.

Producers

In the years before World War II Indonesia produced 80 per cent of the world's pepper. Supplies were cut off during the Japanese occupation resulting in very high prices in the period just after the war. India is now the largest exporter, followed by Indonesia and Sarawak, with the USA being the largest importer.

Diseases

The most serious disease of the pepper vine is foot rot (*Phytophthora palmivora*) which results, initially in a slight droop of the plant, followed by yellowing and then leaf fall, after which the vine dies. One agent in the transmission of foot rot is the introduced giant African snail (*Achatina fulica*) Some rot-resistant varieties are known. Serious pests include bugs, seed-boring weevils and the root-knot nematode (*Meloidogyne javanica*).

Other Cultivated Peppers

The javan long (*Piper retrofractum*) and the betel pepper (*P. betle*) are also cultivated in Indonesia. The former is native to western Indonesia and has small fruits embedded in a fleshy spike, the whole of which is used as a spice. The conservation office of East Java is promoting its growth in the buffer zone around Meru Betiri National Park to increase villagers' incomes and reduce pressure on the forest resources.

The betel pepper is well known for its leaves which are chewed together with a smear of lime and the sliced nuts of the betel palm (*Areca catechu*) sometimes with spices such as cloves or cinnamon. These plants are native to Indonesia but both they and their habit have spread widely: west to east Africa and India and east to the Pacific islands. Betel chewing, an ancient habit, was mentioned by the ancient historian Herodotus in 340 BC. Its purpose is to sweeten the breath after a meal and as a group activity at ceremonial occasions or with friends.

A pepper plantation in East Kalimantan, where the vines grow on wooden supports instead of along the ground under natural conditions. In its natural habitat on the dense forest floor, the vines grow in decaying leaf-mould or humus.

Pepper harvest, from La Cosmographic Universalle, by Andre Thevet, Paris (1575), which is an early, and inaccurate portrayal of the pepper vine as a tree.

Sugarcane and Tobacco

Sugarcane and tobacco plants give rise to two of the most widely used products today. In Indonesia, both crops are grown by smallholders and in large estates in relatively seasonal areas. At present, both are still much in demand but this is likely to decline as the health hazards of their products become more widely appreciated.

Crushing sugarcane on a make-shift mill in Maninjau, Sumatra. The juice which is collected and processed.

»»Plantation workers taking a break atop the load packed for transporting to the processing plant.

A 19th century sugar factory in Surakarta, Central Java, showing the early industrialisation of sugar production.

Growing Sugarcane

Sugarcane plants are cultivars of different hybrids of *Saccharum* species, many of which are indigenous to Indonesia. Sugarcane is a giant grass which grows in dense clumps and may reach nearly two metres in height, with some cultivars having stems up to ten centimetres in diameter. The stem looks superficially like bamboo but is solid and comprises a tough rind around a very fibrous centre. These fibres hold the sweet cane juice which, when chewed and sucked straight from the plant, has a distinctive and pleasant flavour. Sugarcane is the world's major source of sugar, and the term 'cane sugar' has come to be used for all forms of sucrose, irrespective of their origin.

Sugarcane requires high temperatures, plenty of sunlight, good drainage, and a highly fertile soil or generous applications of fertiliser. It also needs an abundant supply of water provided either by rainfall, which must exceed 1,500 millimetres annually, or by irrigation. Sugarcane plants suffer from stem borers (moth larvae); the tunnels they make in the stems provide access for disease organisms such as red rot fungus. Rats also eat the cane and promote the access of disease. A range of snakes live in cane fields, which helps to control the rat population.

Extracting the Sugar

The first stage in harvesting is to set fire to the dried leaves which would otherwise have to be stripped off. The stems are then cut close to the ground because the lower part of the stem is the richest in sugar. Trimmed canes are loaded onto trucks and taken to processing factories where mechanical crushers squeeze out the juice. In traditional processing the juice is boiled down to make thick brown syrup which begins to crystallise, forming dark brown blocks of sugar. Some refinement is necessary for sugar entering the international market because brown sugar formed simply by boiling contains a cocktail of proteins, other sugars, various salts and suspended particles. The albuminous proteins are liable to absorb water from the atmosphere, turning the sugar block into a sticky mess which readily spoils. Modern processing separates the flavourless white sugar from the brown residue known as mollases.

Sugarcane is known to have been used some 3,000 years ago in China, and it was already an important crop in India when it was first introduced to Europe by Alexander the Great in 327 BC, an event which gave rise to tales of reeds producing honey. It was brought to Spain by conquering Arabs in 755

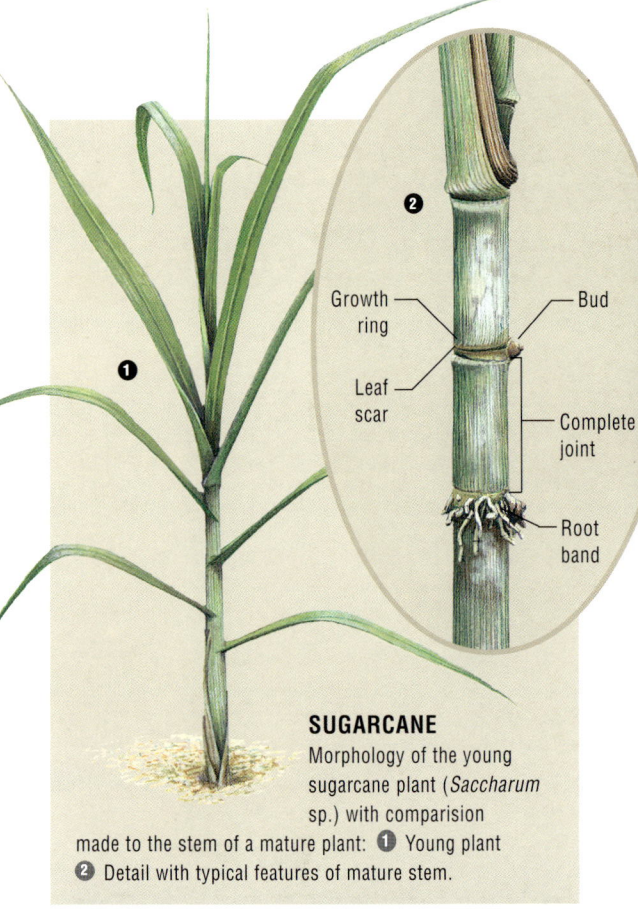

SUGARCANE
Morphology of the young sugarcane plant (*Saccharum* sp.) with comparision made to the stem of a mature plant: ❶ Young plant ❷ Detail with typical features of mature stem.

AD. and subsequently became dispersed throughout the tropical and sub-tropical parts of the Old World. Columbus took some sugarcane to America on his second voyage in 1493, and it was established in the Caribbean by the 16th century.

In areas where it was not grown, sugarcane was initially an expensive luxury food or medicine, replacing honey as the traditional sweetener. The development of sugarcane plantations tended by slaves or bonded labour under colonial powers finally made cane sugar into a widely affordable commodity about 100 years ago. Sugar pressed from canes was produced in Java as long ago as 1635; before this, palm sugar would have been the major sweetener. By the mid-19th century, sugarcane had become an enormously profitable crop for the Dutch and, by the 1920s and early 1930s, sugar was Indonesia's most valuable export.

Total Indonesian production of sugar from all estates and smallholdings in 1993 reached 2.5 million tonnes (52 per cent of this from East Java), but this was not sufficient to meet the domestic demand which is a little over 3.2 million tonnes per annum. Cane sugar has been implicated in a wide range of modern medical disorders and the search is on for less harmful alternatives. Trials of the daisy-like South American stevia shrub (*Stevia rebaudiana*) are promising: its leaves produce crystals which are 200 to 300 times sweeter than cane sugar, but contain far fewer calories.

The History of Tobacco

Tobacco (*Nicotiana tabacum*) is grown for its leaves which, after curing, produce a habit-forming narcotic which can be breathed into the body while the leaf burns. One of the world's most widely distributed crops, it is found from the equator to central Sweden and southern Australia. It does, however, grow fastest in the tropics. This wide distribution belies tobacco's rather specific demands for sunlight, rain, seasons, humidity and soil conditions (different types of soil affect the flavour of the leaf). The most important part of tobacco cultivation, however, is the curing of the leaves which must be carried out in a prepared environment with strictly controlled temperatures and humidity levels. Tobacco suffers from leaf-eating pests which obviously reduce the value of the product. It also faces subterranean adversaries such as mole crickets and the root knot eelworm. The threat of attack from the latter is the reason why tobacco crops are never planted in the same field for more than two consecutive years and also why an interval of seven years is allowed to pass before it is planted there again.

The smoking of tobacco was first observed by Europeans when Columbus reached the West Indies in 1492. At that time its use was widespread in Central and South America, but it was almost a century before tobacco smoking became popular in Europe and tobacco was introduced into Africa and Asia. In Europe, its popularity caused a puritanical backlash, and rulers tried to suppress and control its use, not least by imposing taxes, a major source of national income which remains to this day.

In Indonesia, tobacco is grown in the drier and more seasonal areas such as Northeast Sumatra, eastern Java, and southern Sulawesi. Although tobacco was cultivated locally in the 17th and 18th centuries, it was not until 1863 that the tobacco industry truly began. In that year a Dutch planter moved to Deli, the area around Medan, and established the first commercial tobacco plantation here. The rich volcanic soils of the region produced what soon became known as the world's very best leaf for wrapping cigars. The consequent influx of capital resulted in the first great clearances of Sumatra's lowland forests. By 1890, however, the tobacco boom had reached its height and planters began investing in other crops.

The Ill-Effects of Tobacco

Media campaigns make it clear that cigarette smoking is a major cause of preventable life-threatening disease such as lung cancer, coronary heart disease and pulmonary obstruction. Both smokers and non-smokers exposed to environmental tobacco smoke are at risk, and stopping smoking rapidly reduces those risks. Although the addicting properties of nicotine can make cessation difficult, both medical interventions to help smokers quit and social policies aimed at control of cigarette smoking can have substantial benefits. The nation benefits from smoking because of the taxes levied on the sale of cigarettes and the major employment the plantations and factories provide. Alternative sources of revenue and employment need to be found if this is not to dampen official efforts to discourage the habit.

The various aspects of the tobacco trade. Clockwise: The cultivation of tobacco in Sumatra, from The Graphic *1890. Sorting tobacco for the market and transporting tobacco, from the* Indonesian Photo Posters, *Jean Demmeni.*

*The tobacco plant (*Nicotiana tabacum*) grown for the commercial harvest of their leaves.*

Coffee, Tea and Cocoa

Coffee, tea and cocoa are grown in plantations to produce the raw material for drinks which are popular throughout the world. None of these crops is indigenous to Indonesia, but they are now important commercially, and are extensively grown in certain parts of the country.

Coffee (Coffea arabica) which originated from tropical Africa is now widely cultivated in Sumatra and Sulawesi. Above, a fruiting branch with mature berries.

»»Typical landscape of a tea plantation in West Sumatra.

»Worker in tea plantation with her harvest in her basket, picking tea leaves.

Collection of harvested coffee berries ready for further processing in East Java.

PRODUCTION BY TYPES OF CROPS OF SMALL-HOLDERS AND LARGE ESTATES (1989 – 1993).

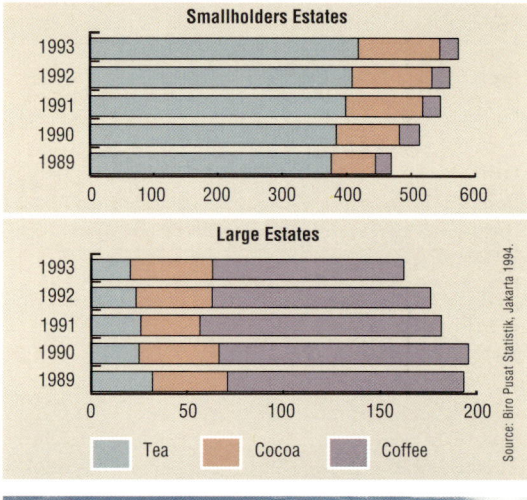

Source: Biro Pusat Statistik, Jakarta 1994.

Tea — Cocoa — Coffee

Coffee Plants from Africa

The taxonomy of coffee is somewhat confused but it has generally been agreed that there are two commonly planted species: arabica *(Coffea arabica)* and robusta *(C. canephora)*. The former is generally held to have the better flavour and dominates the world market, but the latter has superior disease resistance, heavier yields, and fewer growing problems; it is also more commonly grown in Indonesia. Coffee is grown throughout Indonesia in smallholder plots and home gardens. In some provinces it is a very minor product, while in others, such as East Timor, it constitutes the major export commodity.

The Origins of Coffee

Arabica coffee came originally from mountainous areas of Ethiopia, whereas robusta is found from the equatorial forests of West Africa east to Uganda. The first coffee shops were opened in Cairo in about 1550; shops in Paris and London followed suit about a century later, after which coffee became tremendously popular. By the end of the 17th century coffee plantations had been established in Indonesia. Coffee seeds were taken by the French to the West Indies from where it was taken to Brazil, the world's largest producer, in 1727.

In its natural state, coffee is an understorey shrub and is generally grown in dappled shade. Recent experience in Central America, and to a lesser extent in Indonesia, has shown that the tree cover is not essential, although this method is less favourable to the environment. It has clusters of jasmine–scented white flowers which flower at the same time — a behaviour which obviously increases the chances of cross–pollination. The berries, which contains the beans, are dark red when ripe. The fruit pulp is quite sweet and is savoured by monkeys and civet cats. Indeed the best coffee is said to be made from beans found in the excrement of civet cats, since these animals are thought to select only the best berries.

Preparing the Beans

To prepare coffee for the market, the fruit covering and parchment enveloping the bean have to first be removed.

The characteristic flavours develop as the beans are roasted, although the delicious smell of freshly roasted coffee can never be wholly caught in the drink because it is caused by highly volatile oils. The length of roasting is critical; too little and the coffee flavour does not develop, too long and the coffee becomes bitter and the volatile oils are lost.

Tea Plantations from China

Tea *(Camellia sinensis)* is a medium–sized tree native to southwest China and neighbouring regions. Tea leaves have been used as a medicinal beverage in China for almost 3,000 years, and in Japan for nearly as long. The manner in which tea is savoured varies considerably according to custom and personal tastes. The tea might be green, for which the leaves are simply steamed and dried, or black, for which they are withered, rolled, fermented and dried. Black tea accounts for most of the world's tea supply, most of it being produced in India and Sri Lanka. Although tea did not reach Europe until after 1650 and had to compete with coffee, Britain is now the world's major tea consumer by a wide margin. It was not commonly drunk until the 18th century and in Britain it displaced coffee as the favourite drink. As a result, the habit of drinking tea and establishing plantations for its production spread throughout its empire.

Tea grows in sub–tropical lowlands and tropical highlands. It needs moderate temperatures and high rainfall humidity almost all year round. Most of the tea in Indonesia is grown in West Java. Since it would be impractical to pluck the leaves from a 15 metre high tree, the plant is trained into a 1.5 metres high bush. The parts of the plant which are harvested are the terminal leaf bud and two to three leaves below it, together with the intervening stalk. These have the highest caffeine and polyphenol content and thus produce the best tea.

Cocoa from the Americas

Cocoa *(Theobroma cacao)* is a small tree from the understorey of western Amazonian forests. It has quite long leaves and is somewhat unusual in that its flowers grow on its trunk and lower branches. This allows larger fruits and seeds to be produced than if the flowers were attached to slender twigs, and these have a better chance of producing a self-supporting seedling in the gloomy undergrowth than would a tiny seed. The fruit 'pod' grows to nearly 30 centimetres in length and 50 grams in weight, and contains 20 to 70 seeds. A healthy tree can produce about 70 pods each year.

Cocoa was cultivated by Meso-americans for about 2,000 years before Europeans discovered it. It is said that in 1512 the Aztec god-king Montezuma welcomed Cortez, the Spanish conquistador, by giving him a drink of what was called 'chocolatt' flavoured with vanilla, hot peppers, and other aromatic spices. The Spaniards' palate preferred the addition of sugar to peppers and, in this form, exported cocoa became a very popular drink for the wealthy classes in Europe.

Growing and Producing Cocoa

Most of Indonesia's cocoa is grown in northern Sumatra. The cocoa pods are attacked by a wide range of insect pests, but the less pesticides are used, the more these pests are controlled by natural means. Monkeys, rats and squirrels also enjoy eating the seeds.

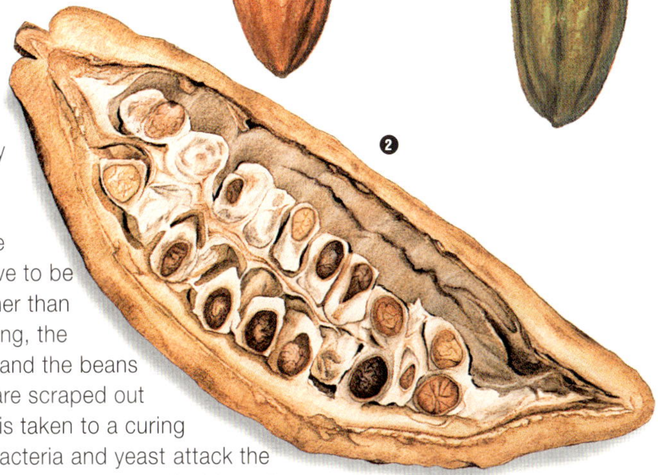

Because of their peculiar position on the plants, cocoa pods have to be harvested by hand rather than by machine. After picking, the pods are broken open and the beans and surrounding pulp are scraped out into a container which is taken to a curing hall for fermentation. Bacteria and yeast attack the sweet mess producing alcohol, carbon dioxide and heat, a process which must not be hurried if the full chocolate flavour is to develop. This, in turn, kills the beans which are then dried in the sun and polished prior to being sent to chocolate manufacturers. The beans then have to be roasted, split open and the kernels ground. The resulting thick liquid, or chocolate liquor, is then used to make cocoa powder or cocoa butter, the latter being used to make chocolate, cosmetics and suntan lotions.

Popular Drug

A common feature of these three beverage crops is that all contain caffeine. This is a mild, tasteless stimulant of the central nervous system which produces increased alertness and capacity for work. The tolerance of individuals to caffeine varies and excessive ingestion can cause sleeplessness, shaking, or stomach upsets.

A cup of arabica coffee typically has 100 milligrammes of caffeine, whereas the harsher tasting robusta has about twice as much. Weak tea has about 35 milligrammes, and a strong cup about double that amount. A cup of cocoa has only about ten milligrammes of caffeine.

Caffeine is removed by placing the raw beans or leaves in very warm water which brings the caffeine to the bean surface. A solvent is than used to remove the caffeine and the beans are then rinsed, dried and roasted.

Tea, a shrub with fragrant white flowers and lanceolate leaves, originally from China, is now grown in Sumatra and Java. The processed leaves are made into an aromatic beverage.

Cocoa (Theobroma cacao), *native to Central and South America is grown in North Sumatra, Central and East Java:*
❶ *Flowering and fruiting branch.*
❷ *Halved pod with numerous seeds.*

Local products made from coffee and tea (from top left): coffee scotch candies, and teh botol or bottled tea.

Oil Palm and Rubber

Oil palm and rubber are two very important plantation crops, covering about one million and 3.5 million hectares of Indonesia respectively, mostly in Sumatra and Kalimantan. They produce important export goods: rubber vies with textiles as the single largest contributor to non-petroleum foreign exchange earnings.

Early oil palm plantations were extremely labour-intensive.

↗*An aerial view of an oil palm plantation in West Sumatra. Notice their characterisitic star-shaped canopies.*

Oil palm fruit — whole and section with indication of position of **A** *palm kernal, an important component.*

Amazonian Origins

The rubber tree *Hevea brasiliensis* is a native of the southern part of the Amazon basin where the local people traditionally used the latex for making balls, bottles, footwear and for waterproofing cloth. In 1770 an English chemist named Priestly discovered that lumps of rubber would rub out pencil marks, and small cubes were later sold for that purpose. More and more uses were found for this product and in 1791, rubber tubing was invented. Some of the early rubber produced for this purpose became tacky in the heat of the sun, but it was then discovered that by heating the latex with sulphur, the rubber would keep its physical properties at normal temperatures.

Rubber plantations were first suggested in 1824 but did not become important until 50 years later when seedlings were taken from Brazil to Asia to support the growing demand for rubber products. After a few unsuccessful attempts at introduction, a group of young seedlings were finally established in Singapore. Removing the trees from their habitat distanced them from the insect pest that attacked the tree in Brazil — the reason this tree is of little commercial importance in its native country.

H.N. Ridley, the first Scientific Director of the Singapore Botanic Gardens experimented with how best to tap the latex, when to tap, and how to control pests and diseases, all with such unbridled enthusiasm and little or no institutional or financial support that he came to be known as 'Mad Ridley'. The first plantation was started in 1889, and his faith in the potential of the tree was vindicated when he saw the great rubber boom of 1910.

While the general image of rubber is of neat and tidy plantations, there are huge areas of 'jungle rubber' grown by small-holders in Sumatra and Kalimantan. This land-use practise begins with forest clearing and non-irrigated rice cultivation among which young rubber trees are planted with a view to future cropping and establishing tenure rights. After a few years of rice cultivation the field is abandoned and left to return to 'bush'. Some six to ten years later, the rubber trees are ready to tap; and this can continue for 20 to 30 years. This jungle rubber is essentially owned as an economically-viable, secondary forest and closely resembles young lowland forest. Its species richness of trees and birds is about half that of primary forest, much greater than a commercial rubber estate.

Oil Palm

The oil palm, *Elaeis guineensis* is a close relative of the coconut *Cocos nucifera* and is native to the rain forests of West Africa, from where it was recorded by a Portuguese voyager in the 15th century. The vernacular name refers to the oil that can be pressed from the red and black fruits. The oil palm was taken to the New World by the slave trade but only became established in Brazil. Plantations have now been developed elsewhere. Four oil palms were received by the Bogor Botanical Gardens in West Java, in 1848 and their offspring were planted all over the archipelago. The last of these original trees survived until just a few years ago. Palms from Bogor and Singapore were later planted in North Sumatra and the progeny from these trees have since been spread throughout Asia. Despite earlier recommendations that the oil palm could be a useful economic crop, it was not until 1898 that the first commercial plantation was established: it was another 20 years before the industry matured. The product was far superior to the oil produced in the villages of Nigeria which was, until then, the main source for world trade.

Oil palms surpass all other crops in the yield of oil per unit area. Two types of oil are produced — one from the fruit flesh, the other from the kernel or seed. Both contain about 50 per cent oil which is used mainly for edible substances such as margarine and cooking oils; kernel oil is used in cakes and confectionery products, ice cream, mayonnaise, toilet soaps, soap powder and detergents. After the kernels have been thoroughly

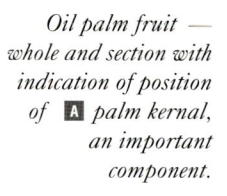

STAGES IN THE PROCESSING OF PALM OIL

The harvested oil palm is transported to the processing plant. After purification, the crude palm oil is refined into its various components and eventually processed into a wide range of end products.

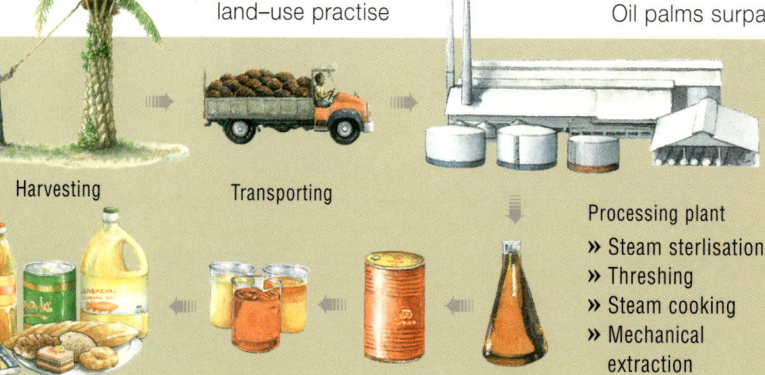

Harvesting — Transporting — Processing plant
» Steam sterlisation
» Threshing
» Steam cooking
» Mechanical extraction of palm oil

Oil palm end products ← Refining processing ← Crude palm oil ← Purification

Recent Palm Oil Developments

There has been a vocal (and sometimes interest-led) nutrition lobby arguing that palm oil is unhealthy because of its high level of saturated fats. Mono-unsaturated fats (such as those in olive oil) have gained prominence as a preferred fat in the human diet. The fat composition of palm oil is, however, markedly different from palm kernel oil or coconut oil, both of which are highly saturated. In fact, palm oil actually raises blood cholesterol levels no more than olive, canola, or peanut oils.

Another threat to the export value of palm oil comes from a genetically-modified rape seed plant recently harvested for the first time by American farmers. This produces oil high in lauric acid, similar to that from palm kernels and coconuts. If the crop is judged a success and is grown more widely to reduce the current 600,000 tonnes of oils imported into the US every year for the manufacture of soaps detergents and cosmetics, it could seriously harm the export of oils from Indonesia and other tropical countries. Meanwhile, research is being conducted on hybridising the common oil palm with a relative, *E. oleifera,* to produce more unsaturated oils.

Wildlife of Plantations

Visits by large animals to plantations, particularly young oil palm plantations, can cause severe economic loss. The main offenders are elephants and forest pigs. In these cases, which are reported almost every month and are a continual threat in certain places, it is pointless to blame the animals. The situation is usually caused by careless land-use planning. If a favoured elephant food (such as young oil palm) is planted next to elephant habitat or across their traditional pathway, they will of course eat the food and there are few effective ways of dissuading elephants from their avowed intention. Options to shield the palms include electric fences and buffer zones of low-grade food plants, though these may attract potential pests.

Bag Worms

In the 1960s and early 1970s DDT and other long-lasting poisons were used to combat pests like bag worms (moth caterpillars) in oil palm plantations. In many cases they survived because the larvae, pupae and adult females were protected by the silk bags woven around themselves. In contrast, the 20 or so species of wasps that laid eggs in bag worms, with the larvae eating it from the inside, had no such protection and died, resulting in an increased bagworm population. The cessation of spraying has meant increased parasitic wasps and a decrease in bag worms.

Clockwise: A worker, assisted by a shoulder torch, tapping rubber in the cool of the night, in East Java. Workers leaving a plantation with their load of tapped rubber in buckets. The processing of the tapped latex into rubber sheets in East Java.

CHARACTERISTICS OF THE 'IDEAL' RUBBER TREE

The 'ideal' rubber tree (far right) is determined by the following parameters: Branches — relatively small, wide angled and perfect articulation, self pruning. Crown — relatively narrow and short. Symmetrical. Healthy foliage. Stem — straight and erect, circular. Rapid girth increment. Renewed bark — smooth and thick. Union, stock/scion — smooth. Virgin bark — smooth and thick. Yield is high per area. The tree should respond well to low frequency tapping and stimulation.

1. Latex flow area of the tapping panel. A and B are the horizontal and vertical flow influence respectively.
2. Latex vessel — dissolution of cell walls to form part of lactiferous system.
3. Bark showing cork layer, stones cells and lactiferous tissue.

Timber and Softwood Plantations

Over the past 25 years, Indonesia has become one of the world's main tropical timber producers and the principal source of plywood. The rain forest resource on which this is based is rapidly depleting, and it has been predicted that by the year 2000 there will be insufficient wood to meet projected demands. Future requirements can therefore only be met from plantations of fast-growing trees.

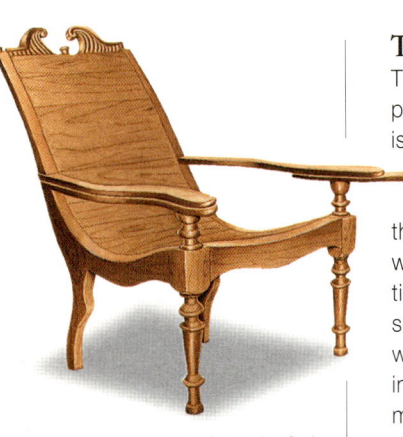

A planter's chair made from the durable teak wood.

Kauri or dammar (Agathis) plantation near Baturaden in Central Java.

Transplanting seedlings onto plots assigned for timber plantation.

DISTRIBUTION OF POTENTIAL INDUSTRIAL WOOD SUPPLY BY MAJOR SOURCE

Source: Ministry of Forestry, Jakarta, 1990.

Natural forest — Plantation forest — Non-forest sources

The Only Route

The rain forest climate zones of Indonesia are perfect for the growth of trees. All year round there is abundant sunlight and moisture. To date the timber industry has been mainly based on natural forests, exploiting the accumulated growth of a century or more; this will continue for a few more decades before valuable timber resources are exhausted. There are recent signs that improvement in natural forest management will ensure limited supplies into the future. This industry, can continue indefinitely into the future as a major contributor to national export earnings only if it adapts itself to using plantation-grown trees for sawn wood, veneer and plywood. By this means, exploiting the natural advantages of the climate, Indonesia can remain a major global timber producer.

Teak and Beyond

Timber plantations in Indonesia began in 1880 with the planting of teak (*Tectona grandis*) in Java. Teak is a valuable timber tree which had been introduced from India many centuries earlier. Since 1880, some 0.9 million hectares have been planted. Like other high-value timber tree crops, teak is grown on a long rotation of 25 to 80 years. At present about 7,000 to 24,000 hectares of teak are being planted every year.

The second most extensively grown plantation species is the native Sumatran pine (*Pinus merkusii*), which has been planted in Java and Sumatra since 1916, mainly for its resin and not for its timber. There are now over 700,000 hectares of plantations. Other major plantation trees are the native *kauri* or damar (*Agathis dammara*), which grows naturally in forests from Sulawesi eastwards and is planted mainly in Central Java, and species of *Dalbergia* and *Pterocarpus,* which are planted for their excellent cabinet-making woods. In addition, there are smaller areas of other species, about 0.5 million hectares. The total extent of these slow-growing timber tree species, all of which are suitable for both sawlogs and veneer, is about 1.5 million hectares. Not all are on good sites, and not all are fully stocked.

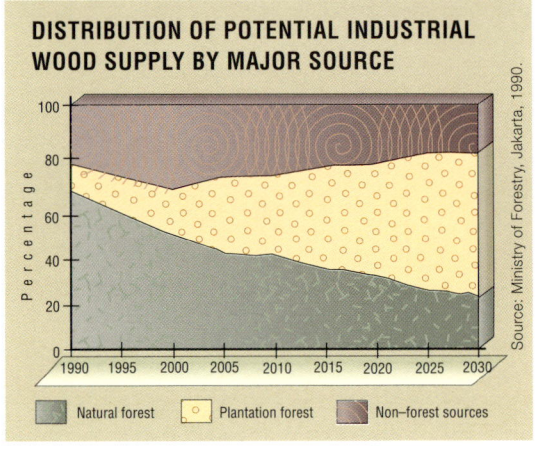

The extensive plantations and forests of pine have important potential as a timber source and not just for resin production. In the mid-1980s, mobile sawmills were introduced to Java to produce sawn pine timber for the domestic market, so that more of the extremely valuable teak timber could be exported.

Fast Growers

Plantations of fast-growing species that reach maturity in eight to 12 years have a much shorter history in Indonesia. Legal encouragement was given in 1984 with a policy to develop industrial plantations. About 150,000 hectares have so far been established and this total is rising. The principal trees are various species of *Eucalyptus* and *Acacia mangium*, some of which are native to eastern Indonesia. These fast growers have been planted by a few private companies mainly in Sumatra, to provide pulpwood for paper mills. Fast growers also produce sawlogs and veneer logs, but tree breeding research needs to be undertaken to improve trunk form. Also, the timber industry needs to adjust to this new raw material — very different from the rainforest logs it has used up to now.

Critical Lands

Besides timber plantations, some trees are extensively planted to rehabilitate critical lands. These critical lands are often watersheds that have become deforested and eroded, and occur on both public and private land wherever there is heavy population pressure, which is mainly on Java. Central government funds are available for this kind of

PEOPLE AND PLANTS: CULTIVATION

*Clockwise: Teak (*Tectona grandis*) trees in the dry season, on an East Javanese plantation where they are often planted in the 'tumpangsari system'. This means teak is usually planted with agricultural crops.*

A tree nursery in Kalimantan where young seedlings are nurtured for industrial forest plantation.

Mature Albizia *sp. in an industrial timber plantation in Kalimantan.*

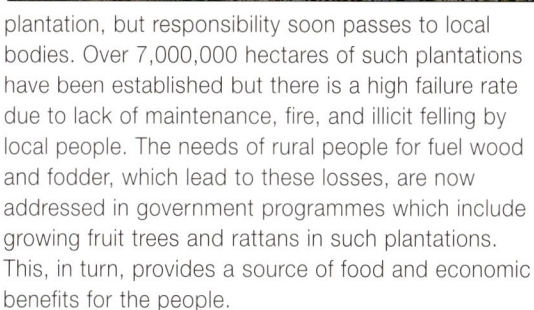

plantation, but responsibility soon passes to local bodies. Over 7,000,000 hectares of such plantations have been established but there is a high failure rate due to lack of maintenance, fire, and illicit felling by local people. The needs of rural people for fuel wood and fodder, which lead to these losses, are now addressed in government programmes which include growing fruit trees and rattans in such plantations. This, in turn, provides a source of food and economic benefits for the people.

Growing Rattan

Over 15,000 hectares of rattan plantations have been established to replace the depleted natural forest sources. Rattan plantations had previously only existed in the Barito valley, Kalimantan, where the local people have been cultivating *Calamus trachycoleus* for over a century. The recent plantations are linked to factories which manufacture furniture for export and the domestic market. Certain species, notably the thick *manau*, are not easy to grow in plantations. Over-harvesting of these rattans has led to a serious reduction in their availability and distribution. They have had to be replaced with moulded or turned timber.

Mangroves

Mangrove forests have been heavily exploited, either for poles or as a source of wood chips for rayon or paper-making. Some mangrove areas have been replanted, the total area covering some 9,000 hectares. Some of these plantations are linked to social forestry programmes which encourage the cultivation of fisheries among the young mangrove trees.

The Future

Future timber harvesting programmes will have to make certain changes to their way of operating, because plantation-grown trees are much smaller than the forest giants still in use today. Advancement in modern technology may allow small billets of wood to be jointed and glued in the near future. During the 1980s the timber of rubber trees (*Hevea brasiliensis*) came to be exploited, using this technology, mainly in Malaysia. In Indonesia 100,000 to 200,000 hectares of old rubber plantations are felled annually, and here too rubber wood need no longer go to waste. The new technology enables small trees to produce high value timber, so increasing the number of species that can be grown in plantations, and reducing the time a plantation takes to reach commercial maturity.

The species-rich forests of Indonesia have numerous promising candidate species scarcely yet explored. One of these, *sungkai* (*Peronema canescens*), is a fast-growing weedy tree of open places, which is related to teak, and has an attractive cabinet grade timber.

Tambak in South Sulawesi with planted Rhizophora micronata *trees.* Rhizophora *sp. are used as a source of chipwood.*

Timber from the ramin tree, a term which refers collectively to species belonging to the genus Gonystylus. *These are distributed throughout Malesia, except in Central and East Java and the Lesser Sunda Islands. Indonesia is the most important exporter of ramin, lightweight to moderately heavy hardwood.*

Orchids: an Introduction

The many islands of Indonesia contain one of the richest array of orchids in the world. Estimates suggest that as many as 4,000 species may occur in the Archipelago, with the greatest number present in Irian Jaya (2,000 species). Orchids are found in virtually every type of habitat, ranging from low-lying coastal regions almost to the upper slopes of the highest mountains. In some places they may even form the predominant vegetation.

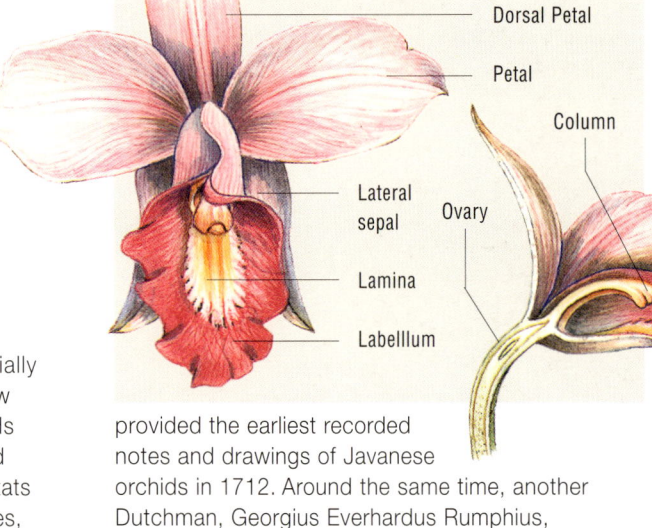

*The leopard orchid (*Grammatophyllum speciosum*), a large epiphyte with distinct petal markings, is found on trees by rivers and in the forest.*

PARTS OF AN ORCHID — Dorsal Petal, Petal, Column, Lateral sepal, Ovary, Lamina, Labelllum

Orchid Distribution

The orchids of western Indonesia are essentially of Southeast Asian origin, while those of New Guinea, Maluku and the Lesser Sunda Islands are Australasian. In Sulawesi, the Asiatic and Australasian orchids intermingle. Not all habitats are equally rich in species — mangroves, grasslands and upper montane areas being relatively species-poor. The richest areas are the hill and montane forests between 500 and 2,000 metres. Forests growing on limestone and serpentine rocks are especially rich in species and are also home to some of Indonesia's most spectacular blooms such as the slipper orchids (*Paphiopedilum* spp.), cymbidiums and moth orchids (*Phalaenopsis* spp.). Riverside forests are another species-rich habitat, and aerial gardens of orchids, rhododendrons, ant-plants and other epiphytic species growing in trees overhanging above the water are a common sight.

Earliest Records

Following his travels in Indonesia during the late 17th century, Engelbert Kaempfer, an employee of the Dutch East India Company in Indonesia, provided the earliest recorded notes and drawings of Javanese orchids in 1712. Around the same time, another Dutchman, Georgius Everhardus Rumphius, spent years collecting, identifying, describing and illustrating the indigenous flora of the island of Ambon in Maluku.

A Diversity of Form, Habit and Flowers

Orchids are amazingly diverse plants in terms of their size and shape. In Indonesia they range in size from the tiny *Bulbophyllum* orchids, scarcely 5 millimetres long with flowers of 2 or 3 millimetres across, to forest giants such as the leopard orchid (*Grammatophyllum speciosum*), which can weigh hundreds of kilogrammes. Climbing vines of the genus *Vanilla* can reach up to 30 metres in length. Vanilla essence comes from the widely grown *V. planifolia* which is indigenous to Central America.

Orchids are equally diverse in their habit. About a quarter of all Indonesian species are terrestrial, growing either on the forest floor or in grasslands and swamps. The majority, however, grow on trees as epiphytes or, less often, on rocks. Epiphytic orchids can be found on the trunks, large boughs or even twigs of trees, depending on the species: a single forest tree may often have as many as 30 or 40 different species growing on it.

The epiphytic environment can be a demanding one with orchids having to survive both torrential rain and desiccating sunshine. Many orchids have evolved special adaptations to conserve water, such as fleshy stems, called pseudobulbs, fleshy or leathery leaves, and thick aerial roots covered in layers of protective, dead cells.

The flowers of orchids are also remarkably diverse. All orchids have relatively simple flowers comprising three sepals, three petals and a central column containing the male and female sexual organs. One of the petals, called the lip, is usually adapted to form a more complex organ bearing callosities, ridges and sometimes, a nectar-bearing spur. The central column is also unusual, in that it normally has only one terminal anther bearing two, four, six or eight pollen masses. The stigma is often a cavity on the ventral surface below the anther but may occasionally be stalked. These adaptations are

IRIAN JAYA'S DENDROBIUMS

Some of the most beautiful of Indonesia's orchids are the dwarf *Dendrobiums* which are found in the highlands of Irian Jaya. These are often tiny, bird-pollinated plants which bear masses of brightly coloured, large flowers in all shades of red, purple, blue and yellow. The best known is *Dendrobium cuthbertsonii* (top left) which usually has red or orange-red flowers. Purple, yellow and bicoloured varieties are found occasionally, and these are highly prized in cultivation. Another spectacular species is *D. vexillarius*, (top right), which commonly bears scarlet flowers. This group of orchids is acquiring a growing band of admirers in Europe and North America and plants are available from nursery-raised seeds.

PEOPLE AND PLANTS: CULTIVATION

related to pollination, which in Indonesian orchids is usually carried out by insects such as bees, wasps and flies. A number of species, however, are pollinated by birds. This is particularly true of certain scarlet-flowered varieties that occur in Irian Jaya.

Strange Biology

Orchids have a complex life cycle which involves interactions with other organisms. Some orchid flowers are extraordinarily well adapted to ensure pollination by the right insect or bird. The flower of *Plocoglottis lowii*, for example, has a furry lip, resembling a bee. When a bee lands on the lip, a spring-like action is set up, throwing the insect onto the anther where the sticky pollen masses become attached to the insect. Some species of *Dendrobium* mimic attractive rhododendrons to ensure pollination, while the large *Dimorphorchis lowii*, has two sorts of flowers to facilitate its pollination. Carrion-smelling flowers are common to the genus *Bulbophyllum*, as in the case of the Bornean variety, *B. beccarii*.

At the other end of the life cycle, all orchids need a fungal partner for their seeds to germinate successfully and grow. This is because orchid seeds do not have a food store (endosperm) to provide energy during the early stages of development. Infection by the right mycorrhizal fungus, however, provides the embryo with food and nutrients for growth which could not take place without this fungal association. In a few species, such as *Epipogium roseum*, this relationship lasts for the entire life cycle of the plant which only appears above ground when it is in flower. These orchids are usually identified as saprophytes in that their sustenance is gained from dead or decaying organic matter rather than by photosynthesis. The majority of orchids, however, can provide their own nutrients once their roots and green leaves have developed.

Conservation

With such an extremely rich array of orchids, many of which are highly prized by people, it is scarcely surprising that some of them are now very rare. The spectacular nature of many species led to an international trade in wild orchids that dates back to the 1840s. The existence of many indigenous species was seriously threatened until a prohibition on collecting wild orchids was imposed by the Indonesian government. Nevertheless, the fact that so many orchids grow as epiphytes on forest trees has meant that they have continued to disappear from the wild as deforestation spread across the Archipelago. Orchid-rich Borneo, Sumatra and Java have suffered most severely in this respect. In Java, the endemic *Phalaenopsis javanica* is probably extinct in the wild as a result of habitat destruction and the depredations of collectors. Similarly, in Borneo, the rat-tailed orchids, *Paraphalenopsis laycockil, P. serpentilingua* and *P. denevei* are on the verge of extinction in the wild.

Many orchids are fortunately preserved in Indonesia's extensive network of nature reserves. The work of the Kebun Raya (Botanic Garden) at Bogor is also aimed at protecting native species.

All Indonesian orchids, both wild and cultivated, are covered by the Convention on International Trade in Endangered Species (CITES) and need a license for their export.

> **SLIPPER ORCHIDS**
> Slipper orchids of the genus *Paphiopedilum* are probably the most popular group of orchids in cultivation, worldwide. They get their name from the deep pouch-like lip at the base of the flower which traps pollinators, usually flies. Such pollinators can only escape by climbing out beneath the anthers, taking pollen with them in the process. Despite a century or more of hybridising, the species is still widely grown and Indonesia has some of the most desirable species. Until recently the demand for these was met by plants collected from the wild and some species are now very rare. Most *Paphiopedilum* orchids are terrestrial, growing in leaf-litter in primeval montane forest. A few, including the widespread Indonesian species *P. lowii* are epiphytic. Many of the native species have single flowers and beautiful leaves which are marked in a mosaic of dark and pale green. The Sumatran *P. superbiens* (top left) is one of the finest, with its sinister, deep maroon flower. Sumatra and Java are also home to several species that flower over a long period. The Sumatran *P. victoria-regina* (below) can produce more than 30 flowers in succession. The Bornean *P. kolopakingii*, in contrast, may produce up to 14 flowers which open simultaneously. Like all slipper orchids these can last for several weeks.

«««Clockwise: Vanilla pilifera *belongs to the genus* Vanilla *whose vines may reach up to 30 metres high into the forest canopy.*

Paraphalenopsis laycockii *one of the orchids which is nearly extinct, in its natural habitat in Borneo.*

Bulbophyllum lobbii, *an inhabitant of the Gunung Gede region in Java, is recognised by the fine strips on its petals.*

Cultivated Orchids

Orchid growing is a popular hobby in Indonesia. Many species, particularly the spectacular hybrids, are often easy to grow and in the equitable climate of Indonesia flourish with the minimum of attention. The hybrid orchids, particularly Phalaenopsis, Vanda *and* Dendrobium *grow well in the lowlands of Indonesia. They are usually more vigorous in growth and have longer-lasting flowers than their wild ancestors.*

*The moon orchid (*Phalaenopsis amabilis*), Indonesia's official 'enchanting' flower is a prized ornamental.*

↗ Vanda *hybrids are widely used in the development of the tropical cut-flower industry.*

Vanda tricolor *one of the three* Vanda *species found in Java.*

Cultivated Species

Many Indonesian orchids are ornamental in their own right. The beautiful moth orchid, *Phalaenopsis amabilis*, is found throughout the Archipelago and is widely grown. Several other *Phalaenopsis* species are found scattered throughout the islands, the most spectacular being the Bornean *P. gigantea* which has leaves that can reach a metre long and a hanging spray of large pink-purple spotted flowers. The slipper orchids (*Paphiopedilum*) are represented by about 25 species, though many are now becoming very rare in the wild. *Cymbidiums* are found on most islands and are recognised by their long sprays of relatively small flowers, in contrast to their Himalayan cousins. Some of the most important species are the *Vanda* orchids and their relatives. *Vanda dearei* is a creamy-yellow flowers from Borneo and is an important species in breeding. The scorpion orchids (*Arachnis*) and scarlet-flowered *Renanthera* species have also been used to produce striking hybrids with long-lasting flowers. The most famous Bornean orchid is the attractive green and black flowered *Coelogyne pandurata,* locally known as the 'black orchid'. It grows in peat swamp forests, which can also be rich in other orchids.

Blooming Industry

Although orchid breeding started in Europe in the 19th century, it has become an important business in Southeast Asia since the first hybrids were raised in Singapore in the 1930s. Indonesia now has a thriving orchid industry based predominantly in Java, with many nurseries around Jakarta and Surabaya.

The main products of Indonesian orchid nurseries are hybrid *Phalaenopsis, Vanda*, scorpion orchids, cane *Dendrobium* and *Renanthera* plants. The *Phalaenopsis* are probably the most popular of these, although fashions can change rapidly. They have arching sprays of white, pink or golden flowers, often with red speckles or stripes on the petals. The flowers can last for many weeks and well-grown plants can be kept in bloom for months at a time. They thrive in Indonesia's humid, warm climate, which is scarcely surprising as many of the wild ancestral species are native to the Archipelago. Many are now, however, very rare in the wild as they are widely collected by orchid growers. *Vanda*, scorpion and *Renanthera* orchids are closely related and are frequently hybridised by growers. *Vanda* hybrids usually have sprays of large flat flowers, often with broad petals in an array of colours from white, yellow and orange to purple, blue and red. The scorpion orchids have more spidery flowers and are often spotted. *Renanthera* orchids and their hybrids are characterised by scarlet flowers. All are useful plants in Indonesian gardens because they can be planted in the ground exposed to the sun and can form a colourful hedge with the added bonus of a crop of saleable flowers. Many of the ancestral species of these hybrids are still found in the wild in Indonesia, with some of the finest being found in Kalimantan.

Cane *Dendrobium* orchids are the other major commercially important group. The main parents of these orchids are the spectacular *D. biggibum* and *D. stratiotes* from Maluku and Irian Jaya. The former has stems up to a metre tall, bearing sprays of long-lasting vivid purple or, occasionally, white flowers. It hybridises freely with *D. stratiotes* and other antelope-petalled dendrobiums, which are mostly found in the island of New Guinea, providing a range of hybrids with white, golden, green, pink, purple or bluish flowers.

Growing Orchids

Orchids can be raised from seed or cloned in large numbers. An average orchid seed pod contains many thousands of dust-like seeds which can be raised in sterile flasks on a nutrient-growing medium. Seedlings can be potted and placed in the nursery in less than a year from sowing: the first flowers are produced after two years. A grower raising an outstanding hybrid can multiply it by cultivating the

A *Cybidium finlaysonianum*, like other members of the genus found in Indonesia, have long sprays of relatively small flowers in contrast to their Himalayan cousins.
B *Dendrobium stratiotes*
C Cane *Dedrobrobium* (*Dendrobium nobile*)
D *Dendrobium biggibum*
E The Bornean *Phalaenopsis gigantea* is the most spectacular specimen with leaves reaching to a metre long and a hanging spray of large pink–purple spotted flowers.

excised growing tip of the stem on a nutrient medium. By dividing the resulting tissue, thousands of cloned plants can be produced with relative ease. These laboratory methods have brought the price of orchids down to a level where they can be bought relatively cheaply and have allowed orchid growing to become an international hobby, which is no longer the preserve of the rich.

Conservation Threats and the Future

For many years, Indonesia had a thriving export trade in wild–collected native orchid species, many of which are outstandingly beautiful. The scale of this trade seriously threatened some of the more desirable species, such as the slipper and moth orchids, with extinction. The export of wild–collected species is now forbidden. The internal market for cut orchid flowers and pot plants has grown so rapidly that most of the country's cultivated orchids are now sold within Indonesia. Increasing demand and the development of new nurseries suggests that Indonesia will soon rival neighbouring countries as a major worldwide exporter of orchids.

The black orchid (Coelogyne pandurata) is one of the most famous Indonesian orchids, its name being derived from the black markings on the lip.

KEY PRODUCTION PROCESSES OF THE ORCHID INDUSTRY

The dust–like propagules are removed from the excised pods and transfer into the culture bottles.

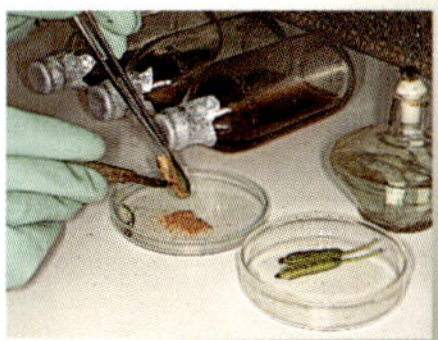

Propagation via the cultivation of seed pods.

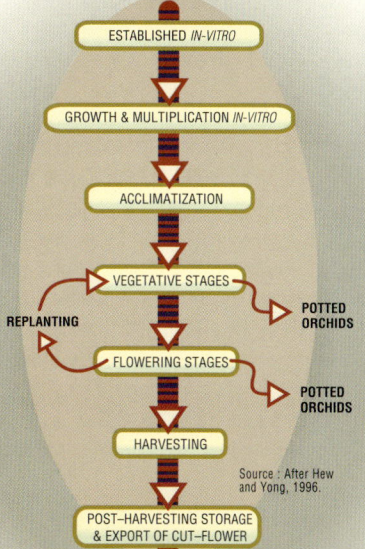

The propagules are grown in a special medium — a mixture of agar, tomato, bean sprout extract and coconut milk.

ESTABLISHED *IN-VITRO*
GROWTH & MULTIPLICATION *IN-VITRO*
ACCLIMATIZATION
VEGETATIVE STAGES → POTTED ORCHIDS
REPLANTING
FLOWERING STAGES → POTTED ORCHIDS
HARVESTING
POST-HARVESTING STORAGE & EXPORT OF CUT-FLOWER

Source: After Hew and Yong, 1996.

The young seedlings are eventually transplanted into individual pots.

Worker removing the seedlings from the medium bottle.

Tranplanting takes place in the greenhouse of the Bogor Botanic Gardens.

Ornamental Plants

Ornamental plants are cultivated all over Indonesia, whether it be in clearings around forest longhouses or the gardens of urban dwellings. Many species such as the sealing wax palm are indigenous, while others, have been introduced to Indonesia during the colonial period when they were used to decorate the spacious gardens of offices and estates, public avenues, and rooms and verandas of houses.

The flame of Irian (Mucuna bennettii) *a climber with spectacular hanging flowers.*

Imports and Hybrids

Some ornamental plants are native to Indonesia. Others, particularly those which traditionally have been used in personal adornment and worship, were probably introduced to Indonesia in ancient times. The majority of ornamental species, however, are relatively recent additions to Indonesia's flora, having been introduced during the colonial period. At that time, they were introduced to decorate public parks and other municipal spaces and to add colour to the spacious gardens of private residences. During the colonial era, hundreds of new plants were brought into cultivation and a great exchange of species took place between continents as professional plant prospectors explored remote parts of the world in order to satisfy the interests and demands of European garden enthusiasts for new and exotic varieties. Tropical plants were even much sought after in Europe where there was a craze for heated conservatories and greenhouses. Gardeners carried seeds and cuttings on their travels and, as a result, many ornamental species now enjoy a worldwide distribution.

New Varieties

In modern times, many new ornamental varieties have arisen by hybridisation and mutation. Such plants are now frequently described as having a secondary origin in addition to their own native origin. Bougainvilleas, for example, are native to Brazil and other parts of tropical America, but the ubiquitous cultivated varieties are descended from hybrids created in Europe by crossing two or more species. These hybrids have, in turn, given rise to mutant forms (or 'sports', as they are referred to by horticulturists) which have arisen spontaneously. In some instances the hybridisation occurred so long ago that it cannot now be determined where, or when, the new species arose. This is the case with the popular *Hibiscus rosa-sinensis*, whose origin is believed to be in the Asia–Pacific region, but this cannot be located or confirmed with any degree of precision.

The easiest way to obtain new varieties is to keep a sharp watch for natural mutations. In some instances, for example, a single variegated shoot may arise from an otherwise normal, green plant. An alert gardener will try to preserve and multiply this sport vegetatively. Looking for sports and propagating them is easy, and it adds to the thrill of discovery to the pleasure of gardening.

»»*A typical, tranquil, Balinese garden, (above), creates a sharp contrast with the bright clashes of colours from the gardens of a modern hotel, (below).*

Hibiscus or 'shoe flower', a highly bred and selected group, used as a decorative ornament on stone carvings in Bali.

PEOPLE AND PLANTS: CULTIVATION

SOME COMMMON ORNAMENTALS IN INDONESIA

1 and **A** *Spider lilies* (Hymenocallis littoralis), *belonging to the herbaceous habit.* **2** and **B** *Bougainvillea* (Bougainvillea spectabilis), *a climbing shrub may reach several meters in height.* **3** and **C** *Hibiscus* (Hibiscus rosa–sinensis), *a common shrub found in gardens and along roads.* **4** and **D** *Javanese ixora* (Ixora javanica), *a medium to tall shrub.* **5** *Sealing wax palm* (Cyrtostachys renda), *an indigenous species threatened by forest clearance and trade as a prized ornamental.*

Collecting and Conserving

New ornamental plants may also be collected from the wild. This practice requires a high degree of ethical responsibility on the part of enthusiasts, collectors and even the gardeners.

Most wild plants are very difficult to cultivate. It should, therefore, be ensured that only seeds and cuttings be collected in this process, without harming the parent plants and without decimating the populations.

BONSAI

The aim in bonsai is to produce a miniature version of a mature tree. The bonsai master will carefully study the form of mature trees, perhaps seeking his inspiration from weather-beaten trees with interesting shapes. Leaves cannot be miniaturized, so small leafed species are a natural choice. The bonsai master starts out by reducing a juvenile tree to the desired size by pruning. The root system is also cut back to fit a suitable container. The branches are then bent and shaped to achieve the desired effect, using wires, weights and fastenings to hold them in position until they have set. Thereafter, new shoots are either selectively removed or else trained in such a way as to maintain the original aesthetic concept. The root system is also regularly pruned and the available space filled with fresh soil to accommodate new growth. One of the key principles of bonsai is to create an internal space for new growth while maintaining the overall dimensions of the tree. In time, mature features such as cracked and fissured bark will develop, which contributes greatly to the character and appearance of the plant.

Making the Most of Weeds

Humans create subjective conditions in which animals and plants are classified according to whether or not they are beneficial — for example if they are edible or distasteful. They commonly become branded as pests if they interfere with human activities or needs. Plants that grow in the wrong places or at the wrong time, competing with crops for resources, or generally creating a nuisance to people, are considered weeds.

The water hyacinth (Eichhornia crassipes) may cause serious problems in rivers, lakes and irrigation canals by spreading rapidly, and covering the entire surface of the water.

Barnyard grass (Echinochloa crus-galli), a weed commonly found in rice fields, with inflorescences above the rice plants.

»»Cyperus rotundus, the purple nutsedge, one of the 'World's Worst Weeds' of food and plantation crops is persistent and difficult to control due to the enormous amounts of tuber produced:
A *Inflorescence*
B *Rhizomes and tuber.*

Survival of the Fittest

Weeds, like all living things, must struggle for their existence. They invade our fields, gardens and other man-made environments where they compete with domesticated species for space, light and nutrients. In the past, efforts were made to eradicate them completely, but, in line with changing attitudes towards the environment, new strategies have been adopted. These were initially aimed at weed control, but more recently have favoured weed management which seeks to keep weed population below a defined threshold of acceptability. Weed management programmes aim to avoid infestation by non-interventionist measures, for example, by instituting quarantine regulations. Herbicides are now only being used as a last resort.

The basic reason why weeds are considered a problem is that they compete with crops for limited resources, the quality and availability of which changes in the course of the crop cycle. Competition can occur both above ground for light, and below ground for water and nutrients from the soil. Crops may develop some advantages through genetically controlled features such as height and the structure of the stem, leaves and roots, or by features such as the relative densities of seedlings and weeds. The optimum density for rice, for example, in terms of obtaining high yields, is also extremely favourable for weeds, which must be crowded out if the crop is to survive.

Weeds, in their turn, exhibit a number of strategies which make them competitive with domesticated plants. An advantage may sometimes be gained by weeds producing substances which inhibit growth in other plants.

More often than not, one finds that weeds have a phenomenal reproductive capacity. They generally produce massive numbers of seed: an approximate 50,000 per plant is a typical quantity in a single year. Such large numbers, however, can only be achieved at the expense of size, and small seeds will only produce small seedlings with a relatively low competitive advantage. Species that have an effective means of vegetative reproduction tend to have fewer, but larger, seeds which typically are capable of dispersal over considerable distances and which posses substantial food reserves that maximise the chances of successful establishment. Plants such as alang-alang (*Imperata cylindrica*) and purple nutsedge (*Cyperus rotundus*) are good examples of this, where the main problem is the production of rhizomes and tubers, respectively, rather than the germination of seeds, relatively few of which are viable. The water hyacinth (*Eichhornia crassipes*) also produces few seeds, although they may remain viable for up to 15 years. This species is also able to reproduce vegetatively by stolons: a single plant can, under favourable conditions, produce 3,000 others in just 50 days and can cover an area of 600 square metres in a year.

Weeds, almost by definition, are very widely distributed. Most Indonesian species are not native to the region but have been introduced, either accidentally or deliberately. Some weed species have seeds, or other parts, which are dispersed by water and enjoy the advantages of irrigated rice fields. Others are dispersed by wind: members of the daisy family and alang-alang, for example, have a pappus, or feathery, umbrella-like structure, which assists their distribution by air currents. Some seeds are dispersed by grazing cattle, while other species, such as Spanish needles

(*Bidens pilosa*), have barbed seed cases which snag onto passing animals. Man is often a major cause of weed dispersal: rice seed is sometimes contaminated with other seeds (though this is not a problem where only the panicles of rice plants are harvested), while seeds mixed with mud may be distributed by farm implements or vehicle wheels.

Noxious Weeds

Weeds are regarded as noxious if they cause serious crop losses or require high management cost. The purple nutsedge, for example, is considered a grave problem among vegetable crops such as onion, chilli, soyabean, groundnut, upland rice and corn, as well as in plantations of sugarcane, oil palm, rubber, coconut and cocoa. The prolific underground tubers can remain dormant and survive extreme conditions of heat, drought and flooding and are extremely difficult to eradicate. Barnyard grass (*Echinochloa crus-galli*), or *jajagoan*, is the most prominent weed species in lowland rice fields, while alang–alang is the most serious problem for upland rice farms and in coconut, rubber, and oil palm plantations. Alang–alang covers between 15 to 30 million hectares of Indonesia and a further 150,000 hectares are invaded annually. Water hyacinth, together with the water fern (*Salvinia molesta*), constitute a serious problem in rivers, lakes and irrigation canals.

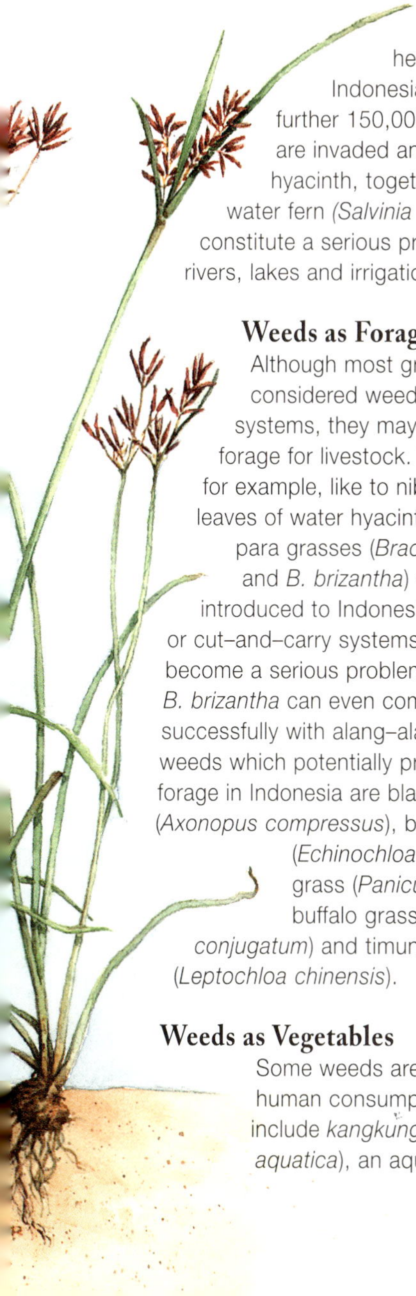

Weeds as Forage

Although most grasses are considered weeds in cropping systems, they may be useful as forage for livestock. Water buffaloes, for example, like to nibble the young leaves of water hyacinth. Two African para grasses (*Brachiaria mutica* and *B. brizantha*) were deliberately introduced to Indonesia for grazing or cut-and-carry systems, but both have become a serious problem in plantations. *B. brizantha* can even compete successfully with alang–alang. Other weeds which potentially provide useful forage in Indonesia are blanket grass (*Axonopus compressus*), barnyard grasses (*Echinochloa* spp.), torpedo grass (*Panicum repens*), buffalo grass (*Paspalum conjugatum*) and timunan grass (*Leptochloa chinensis*).

Weeds as Vegetables

Some weeds are suitable for human consumption. These include *kangkung* (*Ipomoea aquatica*), an aquatic weed that grows in most open waters in the tropics. Another variety is the yellow sawah lettuce (*Limnocharis flava*). The latter grows as a weed in padi fields but is harvested as a vegetable by rural people. *Bayam* (*Amaranthus gracilis*), another widespread weed is also a delicious vegetable, especially when prepared as a salad with peanut sauce. The rhizomes of purple nutsedge may be turned into a Javanese delicacy known as *emping teki*. The roots are boiled, crushed, dried and then fried. *Emping teki* is a local speciality of the Indramayu region of West Java and is commonly seen on sale in the markets of Bandung.

Weeds as Medicinal Plants

Some weeds are considered useful because they are sources of medicines. Corn sow thistle (*Sonchus arvensis*), for example, is used as a diuretic, while bundles of alang–alang rhizomes are sold in markets throughout Indonesia as a cure for headaches or even as an aphrodisiac. *Kangkung* may be taken as a tranquilliser, while a relatively minor weed, *kecubung* (*Datura metel*), is used to treat asthma.

Other Useful Weed Products

Many weeds have considerable potential for various applications. The petioles of water hyacinths, for example, are used as a source of fibre for pulp and paper, and for mats and footwear. Some weeds are being used as a basis of composting materials and, if fermented, as a substrata on which the padi straw mushrooms (*Volvariella volvacea*) can be cultivated.

ALANG–ALANG

The ubiquitous alang–alang grass (*Imperata cylindrica*) provides an interesting example of a plant which is perceived in a variety of ways. In general alang–alang is regarded as an incorrigible weed and large sums of money have been spent on trying to control it. In some areas, however, alang–alang is actually cultivated as a crop which provides thatching materials and fodder for animals. Alang–alang grasslands were once perceived as being far more valuable as a source of thatch and fodder, and as a habitat for game animals, than the forests they replaced. This is reflected in appropriation of alang–alang as a symbol of the aristocracy in Hindu rituals during the pre–Islamic era.

Harvested alang–alang drying in the sun, to be used as thatch in Gunung Kawi, Bali.

Slippers woven from the fibres of water hyacinth petioles which are also suitable material for pulp, shoes and mats.

*Kangkung (*Ipomoea aquatica*), an aquatic weed which is also a palatable vegetable for human consumption.*

Potential for Development of Plant Genetic Variation

Indonesia is one of the world's centres of genetic diversity. These are dominated by the diversity of local fruit, vegetable, some staple crops as well as forest trees and non-forest product plants. The genetic variation of each plant, either cultivated or as a relative, is very important as a gene pool for current plant improvement and for the unpredictable needs of the future. Because of that it is necessary to make an evaluation of their genetic variations.

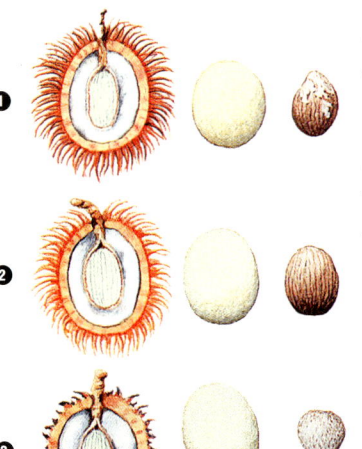

Germ Plasm of the Plant World

Germ plasm is the heredity material transmitted to offsprings through the germ cells, and in each individual, gives rise to the somatic cells. The theory of continuity of germ plasm remaining unchanged from generation to generation, was proposed by August Weismann. DNA is regarded as the molecular equivalent of germ plasm. Germ plasma is seen as biological population, varieties, cultivars, clone, ecotype etc, when 50 samples are taken from the population. The variability of the gene occurs as: 1. today cultivar, 2. previous cultivar, 3. primitive cultivar, 4. cultivar with special characters, 5. wild related species, 6. useful unpropagated species, 7. potential undeveloped species.

Man gathered seeds and fruits of wild plants without significantly altering their nature and it did not lead to domestication until seeds were planted and then harvested. This practice encompassed changes from a. shattering to non-shattering seeds, b. indeterminant to determinant growth, c. unevenly to evenly maturing inflorescence and, d. dormant to nondormant seeds. Further, it led to numerous morphological and structural changes such as increase in the size of seeds, fruits and inflorescence, modification of fruit and inflorescence structure and restoration of fertility in sterile flowers. Traditional genepools provided resistance genes to pests and diseases, for plant breeders all over the world. These evolved in response to limiting factors in the environment, such as biotic, climatic, or edaphic.

Tropical fruits of Southeast Asia are extremely important and have variable resources. Many, such as durian (*Durio zibethinus*), mango (*Mangifera indica*), rambutan (*Nephelium lappaceum*), banana (*Musa paradisiaca*) and langsat (*Lansium domesticum*), are believed to have originated in this region. These are reservoirs of genes which may be used to improve the cultivated taxa. From the horticultural point of view most of these fruits have not been subjected to modern breeding with the farmers retaining traditional methods of cultivation. Propagation done by seeds results in a broad genetic variation within the population. Selection of superior varieties has been conducted among the major groups. Hence, samples of these are maintained in existing agricultural institutes.

Sugarcane is the main plant source for sugar production in Asia, Africa and Madagascar. After an expedition to the Island of New Guinea to collect sugarcane germplasm in 1928, the origin of *Saccharum* species was discovered. Its wild relatives *S. robustum* is found in Indonesia, New Guinea and New Hebrides while *S. sangineum* grows only along the Sepik River of Papua New Guinea. The centre of diversity of sugarcane may have evolved further north in Kalimantan and Sulawesi. Studies show that the maximum diversity of this species is in the Assam-Bengal belt of India. It is likely that the distribution of *Saccharum* was from Japan and Indonesia-New

Rambutan (Nephelium lappaceum), an indigenous species with a wide range of varieties which differ in the thickness, sweetness and juice content of their flesh:
❶ *Variety Binjai*
❷ *Variety Lebak bulus*
❸ *Variety Rapiah*

»»*A cluster of three different varieties of banana (Musa x paradisiaca):*
Ⓐ *P. Raja Bulu*
Ⓑ *P. Barangan*
Ⓒ *P. Susu*

SUGARCANE GERMPLASM EXPLORATION IN INDONESIA

Researchers	Year	Collected clones
Warner and Grassl	1957	S. officinarum (45) S. robustum (7) S. edule (3) Seed collection from S. robustum (5) S. spontaneum (1) and Miscanthus spp (1)
Berding and Koike	1976	Kalimantan (112), Sulawesi (114), Maluku (34) and Irian Jaya (310) – S. officinarum (124) – S. spontaneum (51) – S. robustum (117) – S. edule (24) – Erianthus spp (40)
Tew, Purdy, Lamadji and Irawan	1984	Irian Jaya, Maluku, Sulawesi and North Sumatra – S. officinarum (51) – S. spontaneum (41) – S. robustum (22) – S. edule (6) – Erianthus spp (9) – Miscanthus spp (4)
Sastrowijono	1985	Irian Jaya – S. officinarum (34) – S. spontaneum (48) – S. edule (4) – Erianthus spp (11) – Miscanthus spp (41) – Nephia spp (1) – Unidentified (3)

VAVILOV CENTRES FOR CROP BIODIVERSITY

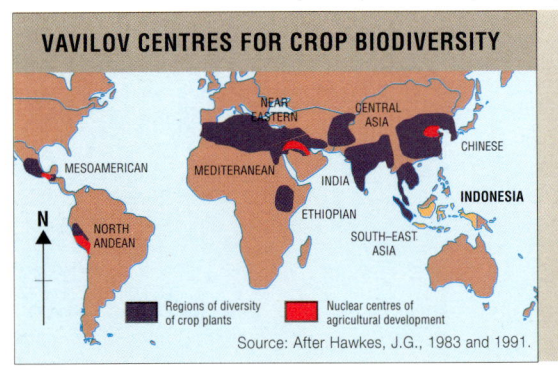

N.I. Vavilov, a Russian botanist first described and mapped centres of diversity for individual crops which he believed represented their points of origin. The centres of diversity of different crops coincided to give remarkable concentration of crop plant variation. Vavilov (1951) ultimately recognised 8 such centres which later authors modified and have also identified new ones (as illustrated at left).

Source: After Hawkes, J.G., 1983 and 1991.

...Guinea, through the Indian subcontinent, to the Mediterranean and Africa.

Primitive Cultivars vs Wild Related Taxa

The value of primitive cultivars in improving the quality of our major food crops has been realised. For tree improvement, however, these values are not properly appreciated. Nevertheless, there is a fear that primitive cultivars of temperate zone fruit trees are disappearing before man will have the chance to make use of them in fruit improvement programmes.

Beside the primitive cultivars, wild taxa are vital for upgrading programmes. However, tropical fruits valued in local markets, such as *lai* (*Durio kutejensis*), *cempedak hutan* (*Artocarpus integer*) and *mata kucing* (*Nephelium malaiense*) are still harvested directly from the forests without any attempt to cultivate them for future harvests.

For improvement programmes, scientists' are usually more interested in using the today cultivars and previous cultivars rather than the wild-related cultivar which has a wide, comparatively unknown, range of genetic variation. With further degradation of forests, similar variations may disappear, resulting in genetic erosion occurring in its natural habitat.

Genetic Variation of Indonesian Plants

Indonesia, home to crops such as banana, nutmeg (*Myristica fragrans*), cloves (*Syzygium aromaticum*), durian and rambutan, has one of the 12 Worldwide Vavilov Centres, for crop biodiversity, it being the largest element in the Indo–Malayan centre. The concept of 'Vavilov Centres' where a centre of current diversity is indicative of the centre of origin of crops, is now considered an oversimplification. However, geographical concentration of crop variation are real and these are of immense conservation importance. The genetic variation of crop plants occurs widely in Indonesia and can be illustrated by the following examples: Banana, one of the more popular fruits, has numerous varieties such as *pisang raja, p. ambon, p.lampung, p. susu, p. raja sereh,* and *p. kapas* belonging to the single species, *Musa paradisiaca*. Each cultivar has it special characteristic which may breed upon other cultivars, such as cultivar *cavendish* to get an improved fruit taste or superior quality as required by consumers.

Durian, one of the favourite domesticated fruits of the region, comprises about 27 species of Durio, although only one is commonly known. A serious disease known as patch canker occurs in the Philippines, Malaysia and Thailand. This necessitates the need for a resistant rootstock to combat the problem. The wild species, generally consumed by the local people in Kalimantan, are still collected from the forests and are still not propagated.

Similarly, *Mangifera indica*, the only known species of Mango, is one of more than 41. A limited breeding programme has been undertaken in Hawaii, but the cultivation of many anthracnose-resistant varieties are required through breeding and selection.

Rice is one of the staples where an advance programme has been implemented. Up to 1994 Indonesia has 7,000 cultivars conserved in Bogor, while 8,248 of *Oryza sativa* and 86 species of its wild relative have been conserved at the International Rice Research Institute in the Philippines. The improvement programme for breeding between 1992 to 1994 has produced more than 100 cultivars some of which are resistant to gall midge, HDB (bacteria found on leaves), DBP (white stripes on leaves) and tolerant to aluminium, iron, acid soil and drought. These have since been released into paddy fields.

The main potential of genetic variation in Indonesia is to improve the plant for agriculture and industrial purposes by breeding. The other areas it shows improvement in include: yield, growth habit, vigour, disease resistance, tolerance to special conditions such as iron, drought, aluminium, acid soil, latex product, plant crown and oil yield.

The improvement of the quality of ornamental plants has only been conducted in selected programmes in which the following are advised: the choice of high yielding varieties for good combination, a range of genetic variability, breeding with wild-relatives species and germ plasm collection for genetic resources.

*Duku or langsat (*Lansium domesticum*), a complex and varied species with a number of wild and cultivated forms (from left):*
A *Variety Duku with loosely presented fruit.*
B *Variety Kokosan with fruits presented in a compact bunch and on a longer raceme than other varieties.*
C *Variety Langsat showing a fruiting branch.*

*A street vendor displaying eggplants (*Solanum melongena*) of varying colours and sizes.*

View of Gunung Kerinci, the highest mountain in Indonesia outside Irian Jaya. It lies within the vicinity of Gunung Kerinci National Park. This is the largest national park in Indonesia and is famous for its immense biodiversity. The park encompasses a wide range of habitats, including lowland dipterocarp forest, montane forest, alpine vegetation, crater lakes and freshwater swamp. In addition, it houses the world's largest flowers — Rafflesia arnoldii *and* Amorphophallus titanum.

CONSERVATION

People have a love–hate relationship with plants. Many cultural groups in Indonesia have a degree of fear, or at least antipathy, towards forests. For others, the beauty or rarity of certain plants are a source of great excitement and pride.

There are those plants we treasure and cosset such as rare orchids, those we encourage and nurture to thrive in our homes, offices and gardens such as sealing wax palms, bougainvillea, yet there are others such as agricultural weeds on which we wage war. In some cases our activities encourage certain species to grow, while in others the conditions we create makes the survival of sensitive species impossible.

The survival of most of Indonesia's flora depends on the continued existence of large areas of natural or near–natural habitat. Most of the conservation areas have been created because of the presence of certain threatened species of animals, but a few small reserves owe their existence to the presence of unusual and rare species. Certain reserves in Sumatra, for example, have been established to protect spectacular *Rafflesia* plants and their vine hosts. That the conservation of plants has been secondary to animals is not, in itself, a problem, since the ecosystems within Indonesia's protected areas system contain much of the country's indigenous flora, as well as its fauna. In some cases though, local endemics, particularly in the lowlands, may be threatened or may have already become extinct because of the widespread effects of logging and land clearance.

Botanic gardens have an important role to play in the general conservation effort. When the giant *Amorphophallus titanum* bloomed in Bogor Botanic Gardens in 1994, there was a constant stream of people who came to gaze on this remarkably wondrous and bizarre sight. Their interest is probably further stimulated by coming into contact with plants which are rarely ever seen in towns and villages. Botanic gardens are limited, however, in lacking the space for more than a handful of individuals of a given species, and so they should never be seen as an alternative to maintaining large protected areas of natural habitat.

Indonesia's flora contains many unknown species, some of which undoubtedly have important medicinal, nutritional or genetic resources. The protection of large areas of natural habitat is the only really effective means of conserving large representations of this country's plants.

The many faces of Indonesian flora echoing the need to preserve its disappearing treasures (from left): A traditional Healer using plant roots from the forest of Sumba. Sandalwood (Santalum album), *prized for its fragrant wood and aromatic oil, is now endangered in its natural habitat. Wild bananas* (Musa sp.) *within a reserve in Java. Roots collected from the forest are used in ikat weaving.*

National Parks and other Major Protected Areas

Indonesia has an excellent record in establishing protected areas covering all major habitat types. With 30 national parks and hundreds of nature reserves already in existence, much of the native flora and fauna has a good chance for survival if the regulations governing the management of these sites is adhered to. Some of the most outstanding areas are described below.

The lush vegetation of the rain forest covered Gunung Leuser National Park in Sumatra which houses more than 4,000 known plant species from Western Malesia.

≫*The banks of the Sekonyer river in Tanjung Puting National Park are lined with freshwater swamp forest. This is also the location of the famous orang utan rehabilitation centre.*

Gunung Leuser National Park (Sumatra)
Gunung Leuser covers 9,500 square kilometres and rises from sea level to 3,404 metres. Because of its large size and range of habitats it is one of Southeast Asia's most important parks, protecting thousands of plant species, some known from nowhere else. Even a short foray into the park will reveal the awe-inspiring diversity of the tropical rain forest, from tall emergents to shade-loving plants of the understorey such as the gingers.

Kerinci Seblat National Park (Sumatra)
Kerinci Seblat offers excellent opportunities to see plants of montane forest and sub-alpine areas, with high-altitude plateaux, the highest freshwater swamp in Sumatra, and mountain peaks rising to 3,800 metres. As Indonesia's largest national park, covering an area of nearly 15,000 square kilometres, it protects the flora and fauna typical of Sumatra, south of Lake Toba. Several of the 26 species of *Rhododendron* found in Sumatra, occur here, including the *Rafflesia* and *Amorphophallus* plants.

Way Kambas National Park (Sumatra)
Although much of the natural vegetation of Way Kambas has been damaged by human activity, the park still contains one of the few areas of lowland dipterocarp forest remaining in Sumatra. Much of the park is marshy, with good examples of non-peaty swamp forest and mangroves.

Tanjung Puting National Park (Kalimantan)
Lying on alluvial land by the coast, Tanjung Puting's highest elevation is just 11 metres. The low-to-medium-sized trees typical of tropical heath forest form the most notable vegetation type, with palms, *Pandanus*, epiphytes and pitcher plants much in evidence. Three blackwater rivers support freshwater swamp forest along their banks, while the stilt or aerial roots and pneumatophores of trees in some areas indicate the oxygen-deficient, anaerobic soil of the peat swamp forest.

Ujung Kulon National Park (Java)
Best known as home to the only viable population of the Javan rhinoceros, Ujung Kulon is interesting floristically because of its recovery from the eruption of Krakatau which sent a devastating tidal wave over much of the peninsula. The lowland forest has many rattans and other palms, and there are excellent areas of coastal forest and mangroves. More interesting still are the four islands of the Krakatau group: three, the shattered remnants

of the original volcano, while the fourth, Anak Krakatau rose from the sea 45 years after the eruption and has provided botanists with a fascinating natural laboratory to study the colonisation pattern of a new island.

Gunung Gede–Pangrango National Park (Java)

The flora of Gunung Gede–Pangrango is among the best studied in Indonesia because of its accessibility and the adjacent Cibodas botanic gardens. Rising to 3,029 metres, the mountain slopes are covered with sub–montane, montane and sub–alpine moss forests, with distinctive stands of Javan edelweiss near the summits.

Baluran National Park (Java)

Located in the drier eastern part of Java, Baluran offers an interesting range of habitats over a relatively small area. Its sparsely-wooded savanna is unique on Java — although it was probably created artificially by fire. Introduced species such as *Acacia* and *Lantana* are causing problems, but the coastal forest, mangroves and the monsoon forest of the hilly slopes are examples of a more original vegetation.

Lore Lindu National Park (Sulawesi)

The principal vegetation type of Lore Lindu is montane forest, with most of the park lying above 1,000 metres. Sulawesi represents the western–most limit of *Eucalyptus* with just one species present, *Eucalyptus deglupta*. This is easily recognisable by its tall, straight trunk and colourful bark whose constant peeling keeps it free of epiphytes.

Morowali Nature Reserve (Sulawesi)

Morowali's significance lies in its extensive alluvial plains (which have largely been converted to agriculture in other parts of Indonesia), and in its ultrabasic soils producing low, scrubby vegetation. Another interesting habitat type is the braided floodplain dotted by *Casuarina* trees. The abundance of pitcher plants growing in the reserve indicate the soil's low mineral content.

Tangkoko Batuangus Nature Reserve (Sulawesi)

North Sulawesi lies along the juncture of two tectonic plates and has several sites of volcanic activity, including a volcano within the Tangkoko Reserve. As a result, the soil is very fertile and supports a high density of fig and other fruit–bearing trees, which in turn support a good population of frugivorous birds and mammals. The altitudinal range is from sea–level to 1,351 metres and habitats include lowland, sub–montane and moss forests.

Ruteng Nature Recreation Park (Lesser Sundas)

This newly created park protects some of the best examples of moist montane forest (deciduous and evergreen) in the Lesser Sundas. It includes Gunung Ranaka which, at 2,140 metres, is the highest peak on Flores. A zone of alpine vegetation exists below the summit of this peak.

Komodo National Park (Lesser Sundas)

Best known for its 'dragons', Komodo's distance from major centres of species distribution compounded by low rainfall has left it with a rather impoverished flora and fauna. The steep, rugged hills are generally covered with coarse grass dotted with lontar palms and *Zizyphus* trees, with thicker stands of trees along water courses and epiphytes such as orchids growing where conditions allow.

Lorentz Nature Reserve (Irian Jaya)

The largest protected area in Indonesia at 21,500 square kilometres, the Lorentz Reserve comprises all major habitat types found in Irian Jaya, including mangroves, swamp forests, montane forests, and sub–alpine and alpine vegetation. The reserve includes Indonesia's highest mountain, the 5,039 metre Gunung Jaya, and an equatorial glacier.

Wasur National Park (Irian Jaya)

Wasur has huge areas of grassland interspersed with reedy marshes and mixed monsoon forest. The most characteristic tree is the paperbark tree (*Melaleuca leucodendron*), whose thick, light bark peels easily and regrows as a strategy to avoid destruction by fire. The park is completely flat and low-lying.

Clockwise: Javanese edelweiss on the top of Gunung Gede in the vicinity of the Gunung Gede–Pangrango National Park. It is reputed to be one of the parks with the best studied flora in Indonesia.

Montane forest in Lore Lindu National Park in Central Sulawesi, located 1,000 metres above sea level.

A view of mangrove and dominant savanna vegetation in Komodo National Park are evidences of the low rainfall of the region.

Giant termite mounds in the open forest of Wasur National Park in Irian Jaya.

The Role of Botanic Gardens

Environmental and social problems are putting increasingly severe pressures on the world's natural vegetation. There are several broad courses of action which can be taken to halt, slow down or even reverse the process. One of these is to increase the number and effectiveness of botanic gardens, particularly in the tropics, where most of the world's plant species are concentrated.

A poster promoting public awareness of the beauty of the flora found in Indonesia's wetlands.

(Below). The main gates of the Botanic Gardens in Bogor.

(Bottom). The gardens in the mountains of Tjibodas are an supplement to the Bogor Botanic Gardens.

Gardens for Plants

The idea of botanical gardens stretches back to antiquity. Gardens in China are known from as long ago as 2,800 BC, while Aristotle is supposed to have begun a botanical garden in Athens in the 4th century BC. These early gardens were primarily concerned with the cultivation of economically useful plants and medicinal herbs. Later the emphasis shifted to the exhibition of ornamental or exotic species, though the scientific dimension never disappeared altogether. More recently, there has been an increasing concern with ecological and conservation issues. This has led to a renewed interest in the role of botanic gardens as scientific institutions.

Botanical Gardens in the Tropics

Many botanical gardens in temperate zones have tropical plants in culture in heated glass houses and have played an important role in the cultivation of tropical species. But nothing can surpass a garden located in the tropics for the finest, natural expression of growth and the greatest ease of culture. A number of outstanding tropical botanical gardens exist today, some of them dating from the 19th century, or even earlier. In Southeast Asia, the largest and best known is the Bogor Botanic Gardens just south of Jakarta. Established by Sir Stamford Raffles in the early 19th century, and continuing today in a long and distinguished tradition, Bogor has made an enormous contribution not just to botany but to agriculture and national progress. It is the largest institution of its kind in Southeast Asia, housing some very rare species.

Other botanical gardens in Indonesia are the mountain gardens of Cibodas in West Java, the seasonal climate gardens at Purwodadi in East Java, and the Eka Karya gardens in central Bali. In addition to these, there are the Grand Forest Parks being developed in every Indonesian province; these have some of the characteristics of botanical gardens.

Modern Conservation Methods in Botanical Gardens

Tropical plants present the conservationist with a number of serious problems — the sheer volume of species being just one. No single botanical garden can hope to grow even one example of every species in its own region or country, let alone maintain a viable self-sustaining population. Another problem is the fact that most of these species are woody and many of these are enormous trees. Furthermore, quite a few plants are tricky to grow in that they are adapted to very specific and often localised ecological conditions. Often their requirements are simply not known, and in this respect, botanical gardens are the ideal place to conduct observations and studies to discover such information. Wherever we look though we see botanical gardens under pressure and in need of public and/or private support to maintain their upkeep, to improve their research facilities, and to employ more and better qualified staff. There is also a need for greater inter-institutional cooperation.

Modern technology plays a vital role in today's botanic gardens. This includes tissue-culture laboratories, automated artificial climate facilities, and sophisticated computer databases. Often, however, there are not enough funds to obtain them.

But it is not only the latest methods and facilities that require funding; there is also the need to maintain the existing infrastructure of botanical gardens: libraries, herbariums, taxonomic research staff, field exploration and seed collection, and training schemes for gardeners and other staff. The larger botanical gardens integrate these activities and are keenly aware of their necessity, but in the world at large

CONSERVATION

«*An avenue of waringins or banyan trees* (Ficus benghalensis). *Handcoloured wood engraving from* Landschappen en Volkstypen van Nederlandsch–Indie, *1883.*

there is sometimes a mistaken notion that only the very latest thing is worth doing. It is, thus, essential that botanical gardens are represented by spokespersons who can explain their vital role to the public, to governments, to educational institutions, and to international agencies.

The Oldest Palm Tree in Asia

Until a few years ago there stood, by the bridge over the Ciliwung River in the Bogor Botanic Gardens, a massive oil palm (*Elaeis guineensis*). This tree, planted in 1848, came from West Africa along with others which met an earlier demise. It was the source of material for experimental cultivation elsewhere and resulted in the millions of oil palms planted across Southeast Asia today.

The Future for Botanical Gardens and Conservation

The need for conservation of the world's flora has never been greater, and the need for botanic gardens runs parallel to this. Existing botanical gardens such as Bogor present enormous challenges and opportunities, but it is no exaggeration to say that the landscape of the 21st century will be partly determined by what can be achieved by such institutions. Indeed, the entire science of botany may be transformed according to the relative success, or failure, of the conservation movement. While it is important to maintain their value as historical institutions as well as their aesthetic appeal, it is essential that the botanical gardens of the world, especially the tropics, commit themselves to the global campaign to protect threatened species and endangered ecosystems. The forests of tomorrow, and their products, will depend on what these institutions can accomplish.

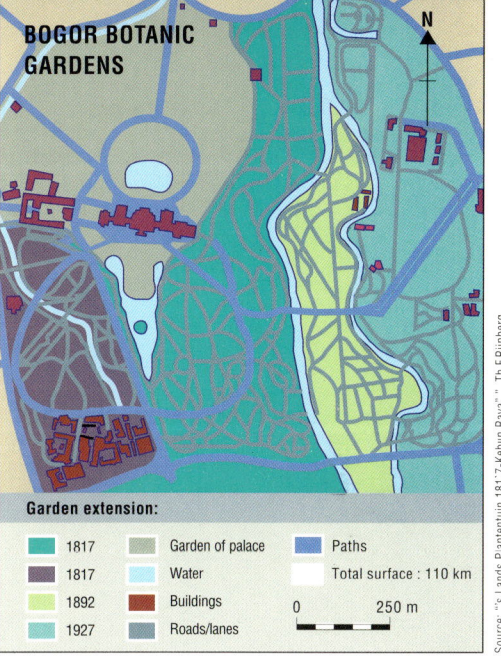

BOGOR BOTANIC GARDENS

Garden extension:
- 1817 — Garden of palace — Paths
- 1817 — Water — Total surface : 110 km
- 1892 — Buildings — 0 — 250 m
- 1927 — Roads/lanes

Source: "'S Lands Pantentuin 1817-'Kebun Raja'.". Th.F.Rijnberg.

⇞ *Aerial view of the northwest corner of the Bogor Botanic Gardens: and the compound of the Presidential Palace which was formerly the site of the official residence of the Dutch governor–generals from 1870 to 1942.*

THE BOGOR BOTANIC GARDENS AS CAPTURED IN A SERIES OF PHOTOGRAPHS BY UNDERWOOD & UNDERWOOD OF AMERICA, 1900 TO 1914.

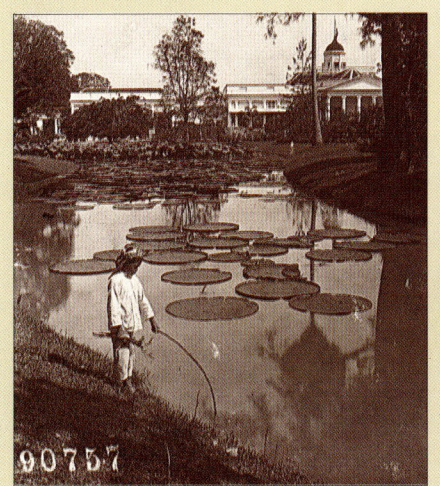

Lake within the compound of the Presidential Palace Garden.

The majestic crowns of the tree ferns which are usually found in the montane forest.

Botanical Gardens' staff members responsible for the daily maintenance of the garden.

135

Glossary

A

adventitious: growth arising from an abnormal position on the plant.

agroforestry: a system of mixed arable forestry and farming.

akar tuba: *(Derris elliptica),* a leguminous climber from which poison is extracted.

alang–alang: *(Imperata cylindrica),* a tall grass.

alpine forest: an area above 3,000 metres that is covered by vegetation types, particular to high altitudes.

alteration of generations: part of a life cycle in which there is a sexual and an asexual phase.

anaerobic: living in the absence of oxygen.

angiosperm: plants that bear seeds.

anther: pollen bearing portion of a stamen.

antheridium: the male reproductive organ of bryophytes, fungi and ferns.

arak: a potent drink produced from the distillation of tuak.

archegonia: the female reproductive organ of bryophytes, fungi and ferns.

aril: the fleshy, often brightly coloured outgrowth at the base of the ovule.

B

***bakau*:** rhizophora.

bract: a small leaf with a relatively undeveloped blade, in axil of which arises a flower or a branch.

buttress: roots growing above ground from the base and trunk to provide support to tall trees growing on shallow or waterlogged soils.

C

callosity: hard, wart–like protective tissue produced in woody plants, often at the site of an injury.

carpel: the female reproductive organ of flowering plants consisting of an ovary, **style** and **stigma**.

cauliflory: the production of flowering shoots on older, thickened leafless branches or on the main trunk.

chlorophyll: the green chemical in plants that is concerned with photosynthesis.

climax forest: the final stage in a succession of forest growth, largely determined by the soil and climate.

corolla: the collective term for the petals of a flower, enclosing the reproductive organs.

cortex: an outer layer of tissue in plants, lying immediately beneath the epidermis.

cotyledon: leaf borne seed on a plant embryo. Two groups are recognised: **monocotyledons** possessing one leaf, and **dicotyledons** with two leaves.

crustose: one of the several growth forms exhibited by **lichens**. It usually forms distinct mosaics on rocks, walls and tree barks.

culm: hollow, jointed stem of a grass or bamboo.

cultivar: a variety of plant which is only found under cultivation.

D

dicotyledon: a plant with two **cotyledons**.

diploid: an organism or cell with double the **haploid** number of chromosomes.

dipterocarpaceae*:* lowland rainforest family of trees found abundantly in Western Malesia.

***dukun*:** a traditional healer who uses plants, herbs and spiritual powers to treat illnesses.

E

ecosystem: a term to describe how organisms interact with one another, and with their environment.

endemic: refers to a species or other taxonomic group that is restricted to a particular geographic location.

emergent: a tree that extends beyond, and produces its own crown above, the canopy layer of a forest.

endosperm: the nutritive matter in seed plant ovules, derived from the embryo sac.

epiphyte: plant not growing parasitically on another plant, but using it as support.

F

filamentous: a growth form of **lichens** which looks like cotton wool, and grows on the bark of trees and shrubs. It is usually orange–yellow or bright green in colour.

foliose: a growth form of **lichen** which is differentiated into an upper and lower surface. It can grow up to 15 to 40 centimetres in diameter.

fruticose: a form of **lichen** which possesses distinctive chemical mechanisms capable of dealing with light intensities. This enables it to grow in the tree canopy.

G

gamete: either of the two germ cells which unite to form a new organism, a mature reproductive cell.

gametophyte: a phase in the life cycle of plants which has **haploid** nuclei. Sex cells are produced during the gametophyte stage of a plant's life.

genome: the set of all different chromosomes found in each nucleus of a given species.

germplasm: hereditary material transmitted to offspring through the germ cells (sexual reproductive cells).

gymnosperm: primitive seed plant in which the seeds are not enclosed in an ovary.

H

haploid: an organism or cell with only one complete set of chromosomes.

haustorium: specialised food absorbing organ, characteristic of certain parasitic plants.

hemi–parasite: an important, and large, group of parasitic plants which consists mainly of mistletoes.

hemi–parasitic: plant species that derive part, but not all, of their nutrients from a host plant.

holdfast: structure in some plants such as algae which serves to attach the **thallus** to a substrata.

hypocotyl: part of a seedling stem below the **cotyledons**.

I

indigenous: species or other taxonomic unit that originates or occurs naturally in a particular geographic region.

inflorescence: flowering shoot of a plant.

IRRI: International Rice Research Institute, an important organisation set up to conduct research into new and less disease-prone varieties of rice.

K

kerangas: heath forest or 'land which will not grow rice'.

L

liana: large woody vine that climbs other plants to reach the forest canopy.

lichen: a fungus (the **mycobiont**) and an algae or cyanobacterium (the **photobiont**) living together in a **symbiotic** relationship.

lodicule: small scale-like structure in grasses beneath the ovary that swells up at the time of flowering, forcing open the **bracts** and exposing the **stamens**.

M

Malesia: the floristic region, on and between the Sunda and Sahul shelves; an area comprising Malaysia, Singapore, Brunei, Papua New Guinea, and the islands of Indonesia and the Philippines.

medulla: central core of tissue in plant stems; functions in food storage.

meiosis: special type of cell division which reduces the chromosome number to half. Occurs in sex cells.

metabolism: the sum of constructive and destructive processes, mainly chemical, that occur in living organisms.

microclimate: the prevailing weather conditions of a small, local area.

mitosis: the usual method of cell division whereby the chromosomes divide in half, each half retained in two new cells resulting from the original cell.

monocotyledon: a plant with only one **cotyledon**.

monopodial: a tree (usually an immature tree) having a single stem with lateral branches.

montane forest: an area with an altitude ranging from above 1,500 to 4,200 metres which is inhabited by an array of plants and vegetation.

mycobiont: the fungal partner in the **lichen symbiosis**.

mycorrhiza: an association of a fungus with the rooting system of a higher plant, in which both species appear to benefit; **mycorrhizal** roots absorb nutrients from the soil far more effectively than similar roots that lack the fungus.

mycorrhizal: pertaining to the non-pathogenic association of a fungus with a vascular plant or bryophyte.

P

petiole: leaf stalk of a plant.

photobiont: The photosynthetic partner in the lichen **symbiosis** which consists of either green algae or cyanobacterium.

photosynthesis: the process whereby plants manufacture complex organic material from carbon dioxide, water and inorganic salts using sunlight as an energy source.

pinnate: leaves in which the leaflets grow opposite each other in pairs on either side of the stem.

pneutomatophore: part of the rooting system of plants that grow in swampy conditions or on **anaerobic** soils. The roots grow upwards and extend above the water.

pollination: in angiosperms, the transfer of pollen grains from an **anther** to the **stigma**. In **gymnosperms**, the transfer of pollen from a pollen-producing cone directly to the ovule.

protonemata: first stage in the development of the **gametophyte** of mosses and certain liverworts.

R

rhizoid: single or several celled hair-like structures serving as a root. Present at the base of moss stems and undersurfaces of liverworts, ferns and some fungi.

rhizome: underground stem bearing buds in axils of reduced scale-like leaves. Serves as a means of vegetative propagation.

runner: see **stolon**.

S

saprophyte: a plant that absorbs soluble organic nutrients from inanimate sources such as dead plants or animal matters.

sepal: one of the parts forming the calyx of **dicotyledon** flowers; usually green and leaf-like.

spadix: an inflorescence enclosed in a **spathe**.

spathe: the **bract**, or pair of bracts that encloses the flower cluster.

spikelet: flower unit that makes up the **inflorescence** of members of the grass family (Gramineae).

spore: a cell in a resistant covering, capable of developing independently into a new individual.

sporophyte: phase of the life cycle of plants which has **diploid** nuclei and during which spores are produced. It arises from the union of sex cells produced by **gametophytes**.

stamen: organ of a flower which produces microspores which are shed after development as pollen grains. The **stamen** is made up of a stalk or filament bearing an **anther** at its apex.

stigma: in a flower, the surface of the **carpel** which receives pollen and on which they germinate.

stolon: stem which grows horizontally along the ground and may form **adventitious** roots which develop into new plants. Eventually these are freed from connection with the parent plant.

style: slender column of tissue that arises from the ovary and through which the pollen tube grows.

symbiosis: term used to describe an association between two or more dissimilar organisms living together in which each member benefits from the others' presence.

sub-alpine forest: a region of forested area which is situated in the higher slopes of mountains but just below the timber line.

T

thallus: simple, vegetative plant body which shows no differentiation between the root, stem and leaf.

tuber: swollen end of underground stem bearing buds in axils of scale-like rudimentary leaves. Tubers contain stored food material and are organs of vegetative reproduction.

V

vascular: pertaining to plants that contain vessels which are differentiated from other cells for the purpose of transporting water, mineral salts and synthesised food materials from one part of the plant to another.

W

Wallace line: an important zoogeographical division firstproposed by the British naturalist and explorer Alfred Russel Wallace to separate the fauna (and to a lesser extent the flora) of the Asian and Australian faunal regions.

Wallacea: the zone of mixing between species of the Oriental and Australian regions, as separated by the **Wallace line**, running east of Bali and Kalimantan.

Bibliography

Anwar J., S. J. Damanik, N. Hisyam and A. J. Whitten. 1984. *Ekologi Ekosistem Sumatera.* Yogyakarta: Gadjah Mada University Press.

Attenborough, D. 1995. *The Private Life of Plants.* London: BBC Books.

Austin, R., D. Ueda and D. Levy. 1970. *Bamboo.* New York: Weatherill.

Backer, C A. 1973. *Atlas of 220 Weeds of Sugar Cane Fields in Java.* Ysel: Deventer.

Backer, C. A., and O. Posthumus. 1939. *De varenflora voor Java.* Buitenzorg: 's Lands Plantentuin.

Backer, C. A. and R. C. Bakhuizen van den Brink. 1963. *Flora Of Java Volumes I – III.* Groningen: Noordhoff.

Beekman, E.M. (ed. and trans.) 1981. *The Poison Tree: Selected Writings of Rumphius on the Natural History of the Indies.* Amhurst: University of Massachussetts Press.

de Beer, J. H. and M. J. McDermott. 1989. *The Economic Value of Non-timber Forest Production in Southeast Asia.* Amsterdam: Netherlands Committee for IUCN.

Bellwood, P. 1985. *Prehistory of the Indo–Malaysian Archipelago.* Sydney: Academic Press.

Brownrigg, H. 1991. *Betel Cutters: from the Samuel Eilenberg Collection.* Stuttgart: Edition Hansjörg Mayer.

BULOG. 1995. *Dalam PJPT I: In the First Phase of Long Term Development.* Jakarta: Badan Urusan Logistik.

Burkill, I. H. 1966. *A Dictionary of the Economic Products of the Malay Peninsula.* Kuala Lumpur: Ministry of Agriculture.

Campbell, D.J. and H. D. Hammond. (eds.) 1989. *Floristic Inventory of Tropical Countries.* New York: New York Botanical Garden.

Campbell, N.A. 1993. *Biology* 3rd ed. California: The Benjamin/Cummings Publishing Company, Inc.

Chandler, R.F. 1979. *Rice in the Tropics.* Boulder, Colorado: Westview Press.

Collins, M. (ed.). 1990. *The Last Rain Forests.* London: Mitchell Beazley.

Collins, N. M., J. A. Sayer and T. C. Whitmore. 1991. *The Conservation Atlas of Tropical Forests: Asia and the Pacific.* London: Macmillan.

Comber, J. B. 1990. *Orchids of Java.* Kew: Bentham–Moxon Trust (Royal Botanical Gardens).

Corner, E. 1952. *Wayside Trees of Malaya,* 2nd ed. Singapore: Government Printing Office.

Corner, E. 1966. *Natural History of Palms.* London: Weidenfeld and Nicholson.

Corner, E. and K. Watanabe. 1969. *Illustrated Guide to Tropical Plants.* Tokyo: Hirokawa Publishing Company.

Covarrubias, M. 1987. *Island of Bali.* 1937. Singapore: Oxford University Press.

Davison, G.W.H. and Chew Yen Fook. 1996. *A Photographic Guide to Birds of Borneo.* London: New Holland (Publishers) Ltd.

Dharma, A.P. 1987. *Indonesian Medicinal Plants.* Jakarta: Balai Pustaka.

van Dillewijn, C. 1952. *Botany of Sugarcane.* Massachussetts: The Chronica Botanica Company, Book Department.

Directorate General of Tourism. 1982. *Indonesia: National Parks and Nature Reserves.* Jakarta: Directorate General of Tourism.

Dransfield, J. 1974. *A Short Guide to Rattans.* Bogor: Biotrop Monograph No. 7.

Dransfield, J. and N. Manokaran (eds.) 1994. *Rattans.* Bogor: PROSEA.

Earl of Cranbrook and D. S. Edwards. 1994. *A Tropical Rainforest: The Nature of Biodiversity in Borneo at Belalong, Brunei.* Singapore: The Royal Geographical Society and Sun Tree Publishing.

Eiseman, F. and M. Eiseman. 1988a. *Flowers of Bali.* Singapore: Periplus Editions.

Eiseman, F. and M. Eiseman. 1988b. *Fruits of Bali.* Berkeley: Periplus Editons.

Everaarts, A. P 1981. *Weeds of Vegetables in the Highlands of Java.* Lembaga Penelitian Hortikultura, Jakarta.

Forbes, H.O. 1989. [1885]. *A Naturalist's Wanderings in the Eastern Archipelago.* Singapore: Oxford University Press.

Hartog, C. D. 1970. 'The Sea–Grasses of the World'. In *Verhandelingen der Koninklijke Nederlandse Akademie van Wetenschappen* deel 59, no. 1. Amsterdam: North–Holland Publishing Company.

Heyne, K. 1987. *Tumbuhan berguna Indonesia.* Jakarta: Yayasan Sarana Wana Jaya.

Hodgson, M., R. Paine and N. Anderson. 1991. *Letts Guide to Orchids of the World.* London: Charles Letts and Co Ltd.

Holttum, R.E. 1954. *Plant Life in Malaya.* Kuala Lumpur: Longman Malaysia.

Hsuan Keng. 1983. *Orders and Families of Malayan Seed Plants.* Singapore: Singapore University Press.

International Rice Research Institute. 1993. *IRRI Rice Almanac.* Manila: International Rice Research Institute.

International Rice Research Institute. 1993. *IRRI 1992–1993: Rice in Crucial Environments.* Manila: International Rice Research Institute.

Jacobs, M. 1970. 'Botanical panorama of the Malesian archipelago (vascular plants)'. In *Natural Resources of Humid Tropical Asia*: 263–94. UNESCO.

Johnsson, D. and A. A. Balkema (eds.) 1991. *Palms for Human Needs in Asia: Palm Utilization and Conservation in India, Indonesia, Malaysia and the Philippines.* Rotterdam: Brookfield.

Khozirah Shaari, A.K. Azizol and M. A. A. Razak (eds.) 1992. *Medicinal Products from Tropical Rain Forests: Proceedings of the Conference.* Kuala Lumpur: Forest Research Institute of Malaysia.

Kok Poh Tin, Hsuan Keng and P. N. Avadhani. 1991. *A Guide to Common Vegetables.* Singapore: Singapore Science Centre.

Lee, D. 1980. *The Sinking Ark, Environmental Problems in Malaysia and Southeast Asia.* Kuala Lumpur: Heinemann Asia.

LBN–LIPI. 1977a. *Ubi–ubian.* Bogor: Puslitbang Biologi – LIPI.

LBN–LIPI. 1977b. *Berberapa jenis bambu.* Bogor: Puslitbang Biologi – LIPI.

LBN–LIPI. 1980. *Jenis rumput dataran rendah.* Bogor: Puslitbang Biologi – LIPI

LBN–LIPI. 1981. *Rumput pegunungan.* Bogor: Puslitbang Biologi – LIPI.

LBN–LIPI. 1984a. *Polong–polongan perdu.* Bogor: Puslitbang Biologi – LIPI.

LBN–LIPI. 1984b. *Kerabat beringin.* Bogor: Puslitbang Biologi – LIPI.

MacKinnon, K.S. 1992. *Nature's treasurehouse: the Wildlife of Indonesia.* Jakarta: Penerbit PT Gramedia Pustaka Utama.

MacKinnon, K. S., G. Hatta, H. Halim, and A. Mangalik. 1996. *The ecology of Kalimantan.* Singapore: Periplus Editions.

MacKinnon, K. S., G. Hatta H. Halim, and A. Mangalik. 1996. *Ekologi Kalimantan.* Jakarta: Periplus Editions.

van der Maessen, L.J. G., and S. Somaatmadja. 1989. *Plant resources of South–East Asia: Pulses.* Wageningen: Pudoc

Mannetje, L.'t. and R. M. Jones. 1992. *Forages.* Bogor: PROSEA.

Maxwell, George. 1907. *In Malay Forests.* Edinburgh and London: William Blackwood and Sons Ltd.

McClure, F. A. 1966. *The Bamboos: A Fresh Perspective.* Cambridge: Harvard University Press.

van Meeuwen, M.S., H. P. Nooteboom and van Steenis, C. G. G. J. 1961. 'Preliminary revisions of some genera of Malaysian Papilionadae I'. In *Reinwardtia* 5: 419–50.

Ministry of National Development Planning. 1993. *Biodiversity: Action Plan for Indonesia.* Jakarta: Ministry of National Development Planning/National Development Planning Agency.

Mitchell, A.W.. 1986. *The Enchanted Canopy: Secrets from the Rainforest Roof.* n.p.: Collins.

Monk, K. A., Y. de Fretes and G. Reksodihardjo–Lilley. 1996. *The Ecology of the Lesser Sundas and Maluku.* Singapore: Periplus Editions.

Monk, K. A., Y. de Fretes, and G. Reksodihardjo–Lilley. 1996. *Ekologi Nusa Tenggara dan Maluku.* Jakarta: Periplus Editions.

Othman Y. and S. Subhadrabandhu. 1995. *The Production of Economic Fruits in South–East Asia.* Singapore: Oxford University Press.

Pancho, J. V., and M. Soerjani. 1978. *Aquatic weeds of Southeast Asia.* Quezon City: Nat'l Publ. Corp.

Petocz, R. G. 1989. *Conservation and Development in Irian Jaya.* Leiden: E.J. Brill.

Pfeiffer, J. n.d. *Wild Orchid: The Indonesian Botanic Garden Collections.* Jakarta: Grafimatra.

Piggott, A. G. and A. C. Piggott. 1988. *Ferns of Malaysia in Colour.* Kuala Lumpur: Tropical Press.

Piper, J.M. 1989. *Fruits of South–East Asia.* Singapore: Oxford University Press.

Piper, J.M. 1992. *Bamboo and Rattan: Traditional Uses and Beliefs.* Singapore: Oxford University Press.

Piper, J.M. 1993. *Rice in South–East Asia: Cultures and Landscapes.* Kuala Lumpur: Oxford University Press.

Polunin, I. 1987. *Plants and Flowers of Singapore.* Singapore: Times Editions.

Polunin, I. 1988. *Plants and Flowers of Malaysia.* Singapore: Times Editions.

Reed, C. A (ed.) 1977. *Origins of Agriculture.* The Hague: Mouton.

Richards, P.W. 1952. *The Tropical Rain Forest.* Cambridge: Cambridge University Press.

Ridley, H.N. 1912. *Spices.* London: Macmillan and Company Limited.

Sharp, I. and A. Compost. 1994. *Green Indonesia: Tropical Forest Encounters.* Kuala Lumpur: Oxford University Press.

Shepard, B. M., A. T. Barrion and J.A. Litsinger. 1987. *Friends of the Rice Farmer: Helpful Insects, Spiders, and Pathogens.* Manila: International Rice Research Institute.

Siemonsma, J.S. and K. Piluek (eds.) 1994. *Vegetables.* Bogor: PROSEA.

Smith, B.D. 1995. *The Emergence of Agriculture.* New York: Scientific American Library.

I. Soerianegara and R.H.M.J. Lemmens. 1994. *Timber Trees: Major Commercial Timbers.* Bogor: PROSEA.

Soerjani, M., A. J. G. Kostermans and G. Tjitrosoepomo. 1987: *Weeds of Rice in Indonesia.* Jakarta: Balai Pustaka.

van Steenis, C. G. G. J. 1972. *The Mountain Flora of Java.* Leiden: E.J. Brill.

van Steenis, C. G. G. J. 1981. *Flora untuk Sekolah di Indonesia.* Jakarta: Pradnya Paramita.

Stone, D. 1994. *Tanah Air.* Singapore: Archipelago Press.

Tan Teck Koon. 1990. *A Guide to Tropical Fungi.* Singapore: Singapore Science Centre.

The State Ministry of Environment and KONPHALINDO. 1995. *An Atlas of Biodiversity in Indonesia.* Jakarta: MNLH and KONPHALINDO.

Veevers–Carter, W. 1984. *Riches of the Rain Forest: An Introduction to the Trees and Fruits of the Indonesian and Malaysian Rain Forests.* Singapore: Oxford University Press.

Veevers–Carter, W. 1986. *A Garden of Eden: Plant Life in South–East Asia.* Kuala Lumpur: Oxford University Press.

Verheij, E. W. M. and R. E. Coronel. 1992. *Edible fruits and nuts.* Bogor: PROSEA.

Wallace, A.R. 1986. [1869]. *The Malay Archipelago.* Singapore: Oxford University Press.

Warren, W., A. J. Whitten, M. Jenkins, A. Vickers and T. Hunter. 1995. *Balinese Gardens.* Singapore: Periplus Editions.

Whitmore, T. C. 1977. *Palms of Malaya.* 2nd ed. Oxford: Oxford University Press.

Whitmore, T. C. 1984. *Tropical Rain Forests of the Far East.* 2nd ed. Oxford: Clarendon.

Whitmore, T. C. 1990. *An Introduction to Tropical Rain Forests.* Oxford: Clarendon.

Whitten A. J., M. Mustafa and G. S. Henderson. 1987. *Ekologi Sulawesi.* Yogykarta: Gadjah Mada University Press.

Whitten, A. J., M. Mustafa and G. S. Henderson. 1987. *The Ecology of Sulawesi.* Yogyakarta: Gadjah Mada University Press.

Whitten, A. J., S. J. Damanik, J. Anwar and N. Hisyam. 1987. *The Ecology of Sumatra.* 2nd ed. Yogyakarta: Gadjah Mada University Press.

Whitten, A. J. and J. E. J. Whitten. 1992. *Wild Indonesia.* London: New Holland.

Whitten, A.J., R. E.Soeriaatmadja and S. Afiff. 1996. *The Ecology of Java and Bali.* Singapore: Periplus Editions.

Whitten, A. J., R. E. Soeriaatmadja and S. Afiff. 1996. *Ekologi Jawa dan Bali.* Jakarta: Periplus Editions.

Wilson, H.K. 1948. *Grain Crops.* New York and London: McGraw Hill Book Company Inc.

Wilson, E.O. 1993. *The Diversity of Life.* London: Penguin.

Yong Hoi Sen. 1981. *Magnificent Plants.* Kuala Lumpur: Tropical Press Sdn. Bhd.

Zohary, D. and M. Hopf. 1988. *Domestication of Plants in the Old World.* Oxford: Clarendon Press.

Index

A

Abrus precatorius 91
Acacia 43, 90, 133
   A. auriculiformis 18
   A. farnesiana 90
   A. leucophloea 45
   A. mangium 77, 118
Acanthus ilicifolius (red sandalwood) 27, 37
Aceh 111
Acer laurinum (Indonesian maple) 39
Achatina fulica (snail) 111
Acriopsis javanica (orchid) 61
Acrostichum aureum 37
Adenanthera pavonina 91
Aethopyga siparaja (crimson sunbird) 50
Agathis sp. *(kauri)* 27, 71, 71, 76, 77
   A. dammara (dammar) 70, 118, 118
agriculture 13
agroforestry 88, 89, 93
air roots 34
akar tuba 69
alang-alang 10, 45, 126, 127, 127
Alangiaccae 64
Albizia sp. (timber species) 76, 119
   A. chinensis 45
   A. saman 91
Albugo ipomoeae-aquaticae (crop disease) 18
Aleurites moluccana (candlenut) 43, 97
algae 6, 9, 20, 21, 39
aloewood 71, 71
alpine forest 40–1, 130, 133
Alpinia (ginger species) 106
   A. galanga (lengkuas) 107
Alstonia pneumatophora 34
Amanita (fungi) 18
Amaranthus gracilis (bayam) 127
Amherstia noblis 91
Amorphophallus sp. 132
   A. decus-silvae 28
   A. paeoniifolius 29
   A. titanum 6, 28–9, 28, 29, 130, 131
Amyema anisomeres (mistletoe) 36
Amylomyces (mould) 95
amylose 86
amyotytic moulds 95
Anak Krakatau 10, 11, 52, 132–3
Anaphalis javanica (Javanese edelweiss) 40
andong 74
Angi Gigi lakes 61
angiosperms 25–6, 26, 27, 76
Anisoptera (mersawa) 76
ani–ani 86
Annonaceae 51
annuals 44
ants 47, 60–1
ant plants 33, 42, 60, 61, 120
Anthoceros sp. (hornwort) 23
Antiaris toxicaria (upas tree) 69
apothecia 21
Aquilaria malaccensis (aloewood) 71
arabica (coffee species) 114
Araceae (aroids) 16, 29
Arachis hypogaea (groundnut) 92
Arachnis spp. (scorpion orchids) 122
arak (palm wine) 102
Araucaria spp. (monkey puzzle) 27, 76
Areca catechu (areca nut) 11, 66, 67, 67, 98, 101, 101, 111

areca nut. *See* Areca catechu
Arenga sp. (palm) 50
   A. pinnata (sugar palm) 98, 102, 102
aroids 16, 29, 88
Artocarpus spp. 96, 129
   Artocarpus heterophyllum (jackfruit) 96
   Artocarpus cempedak (cempedak hutan) 129
artocarpus seeds 97
Aru group 44
arum lilies 29
asam jawa. *See* tamarind
Ascomycete (yeasts) 18, 19
asexual reproduction. *See* reproduction process
Asian elephant 58
Asmat (Irian Jaya) 62, 104, 105
Aspergillus oryzae (Deuteromycete) 19
Asplenium nidus (bird's nest fern) 17, 25, 60
Atoni Pah Meto 66
aubergine 46
Auricularia spp. (ear fungi) 18
Australian heath 40. *Also see* heath forest
Averrhoa carambola (starfruits) 97
Avicennia sp. (mangroves) 36, 37
avocado 97
Axonopus compressus (blanket grass) 127
Azadirachta indica (neem tree) 43

B

bag worms 117
Bahau River 72
bakau 76
Balanophora elongata (root parasite) 16
Bali 86
Bali Barat National Park 42
Baliem Valley 21, 38, 62, 85, 95
balsams 39
Baluran National Park (Java) 42, 133
bamboo 6, 42, 63, 66, 74–5, 88
bambu betung 74, 75
bambu duri 74
bambu gombung 74
bamboo shoots 49, 63, 64, 75
Bambusa. *See* bamboo
Bambusoideae. *See* bamboo
banalu 54
bananas 16, 46, 66, 67, 88, 95, 96, 128, 128, 129, 130
Banda Islands 108
Bangka Island 110
Banjarmasin 96
bans, export 72, 76
Banten 111
Barisan Range 31
Barito valley 119
barnyard grass 127
Barringtonia fruit 52
Barringtonia seedlings 10
Barringtonia acutangula 34
Basidiomycetes 18
basketry 103
bats 51, 52, 52
batai 91
batik industry 101
Batur 11
Bauhinia 90
   B. purpurea (orchid flower tree) 91
bayam 127
beans 26, 92–3
Beccarian bodies 60, 61
bees 50–1, 50
beech trees 38
begonias 88

belukar. *See* secondary forest
bengkowang 93
benor 65, 65
benzoin 71, 88
Berbak National park 34
betel palm 11, 83, 98, 111
   b. chewing 101, 101, 111
   b. leaf 67
   b. nut 101
   b. pepper 101, 111
bidans (midwives) 68
Bidens pilosa (spanish needles) 126
binjai (mango) 96
bintangor 71
biomass 47
bird's nest fern 17, 25, 60
birds of paradise 39
bladderworts 35, 56–7, 56
blanket grass 127
blowpipes 69
Boesenbergia (ginger sp.) 106
bog 40
Bogor Botanic Gardens 29, 35, 116, 121, 123, 131, 134, 134, 135, 135
bole climbers 16
Boletus (fungi) 18
Bombacaceae 64
bonsai 125
Borassus flabellifer (lontar or palmyra palm) 44, 98, 99, 102, 102, 103, 133
boring beetles 109
Bornean Ginalloa 54
Borneo 34, 43, 61, 121
botanical gardens 134–5. *Also see individual listings*
Bouea macrophylla (gandaria) 96
bougainvilleas 124, 125, 131
Brachiaria mutica (grass species) 127
breadfruit 85
brem 95
Bruguiera (mangrove tree) 36, 37, 37
bryophytes 6, 22
Buceros bicornis (greater hornbill) 52
bugs 87, 111
Bukit Kelam 56
Bulbophyllum orchids 120, 121, 121
Burmannia 56
burning, forest 35, 39, 40, 41, 43, 46, 47, 69
Burseraceae 70, 71
Buru Island 42, 43, 44
buttercups 40
butterflies 50, 51, 59

C

cabbages 65
Caesalpinioidea (legumes) 90, 91
Caesalpinia pulcherrima 90
caffeine 115
cajuput oil 70
calamondin 97
Calamus sp. (rattan) 60, 72–3, 73, 98, 99
   C. trachycoleus (rattan) 119
Calliandra calothyrsus (legume) 93
Callosciurus notatus (plantain squirrel) 48–9
Calophyllum sp. 14, 34, 71
Calopogonium mucunoides (calopo) 90
Calymperes sp. 23
Camellia sinensis (tea) 114–5, 114, 115
camphor 64, 70, 71
Cananga odorata (sandat) 67
Canarium spp.
   C. commune (kenari) 105
   C. megalanthum (wild

meritus) 97
   C. indica (kenari) 97
   C. vulgare (kenari) 97
canarium tree 85
Candida (yeast) 95
candlenut 43, 85, 97
cane furniture industry. *See* furniture industry
canopy, forest 12, 14–5, 16, 32, 39, 42, 59
Calophyllum spp. *(bintangor)* 71
Carica papaya (papaya) 97. *Also see* papaya
carnivorous plants 56–7
cash cropping 88–9
cashew 96, 97
cassava 83, 85, 94, 94, 95
Casuarina junghuhniana (she-oak) 11, 39, 43, 50, 45, 133
catfish 53
cauliflory 16
Ceiba pentandra (kapok tree) 89
cempedak hutan (fruit) 129
Cenderawasih 58
Centrosema pubescens (centro) 90
Ceriops sp. (mangroves) 36, 37, 49
Ceylon oak 43
champak 67
chempaka (Ambon flower) 26
chilli pepper 46
chlorophyll 54
Chrysopogon (grasses) 44
Cibodas Botanic Gardens 133, 134
cili (Balinese ritual doll) 7
Ciliwung River 135
Cinnamomum burmanni (cinnamon) 83, 88
Cinnamonum camphora (camphor tree) 71
Citris mitis (calamondin) 97
Citrullus lanatus (watermelon) 97
citrus fruits 83. *Also see* fruits
Citrus maxima (pomelo) 97
Clarias batrachus (common walking catfish) 53
clearing, forest. *See* forest clearance
Clerodendrum 61
climbers 12, 16, 31, 47
cloud forest 23
cloves 83, 109, 109, 129
clubmosses 24, 25
Coccocarpia 20
Coccus spp. 47
cocoa 83, 115, 115
coconut 52, 66, 67, 83, 85, 88, 98, 99, 100–1, 100, 116
Cocos nucifera (coconut). *See* coconut
Coelogyne pandurata (black orchid) 9, 122
coffee 53, 83, 88, 114, 114
Coffea spp. 114
   C. arabica (arabica) 114, 114
   C. canephora (robusta) 114
collision, continental (chart) 10
Colobine 59
Colocasia esculenta (taro) 85, 85
Compositae (daisies) 26, 27
conifers 6, 11, 27, 76
conservation programmes 15, 29, 41, 72–3, 78–9, 80–1, 89, 121
Convention on International Trade in Endangered Species of Fauna and Flora (CITES) 25, 121
conversion forest 80
Cookeina tricholoma (cup fungus) 18, 19

copra 100
Coprinus plicatilis (ink caps) 18
coral tree 91
corn sow thistle 127
Corypha utan (gebang) 98, 99
Costus sp. (ginger) 106, 106
cotyledons 26
co-evolution 59
Crematogaster spp. 47, 60, 61
cucumber 46
cultivated species 83
Curcuma sp. (ginger) 106
  C. longa (turmeric) 106, 106
Cyamposis tetragonolobus (guar) 92–3
cyanobacterium (algae) 20, 21
Cyathea spp. (tree fern) 24, 39
  C. latebrosa (tree fern) 24
Cyatheaceae 25
Cycadopsida (cycads) 6, 27, 27
cymbidiums 120, 122,
Cymbidium finlaysonianum 123
Cynopterus sphinx (fruit bat) 52
Cyperus rotundus (nutsedge) 126, 126, 127
Cyrtostachys renda (sealing wax palm) 99, 124, 125

D

Daemonorops (rattan) 60, 71
daisies 26, 27, 41, 126
Dalbergia (timber species) 118
  D. latifolia (Indian rosewood) 43, 76, 91
dammar 27, 70, 71, 71, 76, 77, 118, 118
damar minyak 71, 76
Dampier, William 45
Danaidae (milkweed butterflies) 59
Danau Sentarum 34
Dani (Irian Jaya) 62, 65, 85, 94
Datura metel 127
Dayak (Kalimantan) 62–3, 64, 69, 70, 72, 85, 75
deciduous forest 42
dedap 66, 91
deep water rice 86, 86
defence systems, types of 58–9
defoliators (pest) 87
deforestation 78–81, 118, 121
Delonix regia (flamboyant tree) 90, 91
dendrobiums 120, 121, 122, 123
Dendrobium nobile 123
Dendrobium taurinum 9
Dendrocalamus 74
  D. asper (bambu betung) 74, 75, 75
Dendrolagus ursinus (Vogelkop tree kangaroo) 58
Dendrophthoe magna 54
Dennstaedtiaceae (weeds) 25
Derris spp. 69
  D. elliptica (akar tuba) 69
Detzneria (alpine shrub) 41
Deuteromycetes (fungi) 19
Dewa Kaki 66
dewa sanga (annual feast) 66
Dewi Sri 66, 66
Dialium (keranji) 91
Dicaeum spp. (flower peckers) 54, 55

Dicerorhinus sumatranesis (Sumatran rhinoceros) 58
dicotyledons 26
Dictyophora indusiata (stinkhorn fungus) 18, 19
digestion inhibitors 58–9
Dimorphorchis lowii (orchid species) 121
Dioscorea spp. (yams) 65, 85
  D. hispida (gedung) 65, 65
  D. piscatorum (tuba hubi) 65
Diospyros (ebony) 16
Dipterocarpaceae 9, 10, 11, 26, 28, 30, 31, 32, 33, 49, 52, 52, 64, 70, 73, 76, 80
Dipterocarpus littoralis (palanar) 71
Dischida sp. (epiphytic ant plant) 42, 60
  D. nummularia 61
  D. rafflesiana 61
diseases, plant 18, 87, 109, 111, 112, 116, 129
dispersal agents
  animals 52, 126
  bird 29, 47, 55
  water 36, 52, 100, 126
  wind 27, 47, 50, 52, 77, 126
distribution 9, 10, 11, 14, 44, 74
forest area (chart) 80
diversity 9, 11, 13, 14, , 26, 32, 36, 80, 81, 89, 96, 98, 128, 129, 130, 132
dragon's blood 71
drifting, continental 10, (diagram) 10–1
Drimys (magnolia) 40
Drosera sp. (sundews) 56, 57, 57
dry forest 42, 102
dry upland rice 86, 86, 87
Dryobalanop (kapur) 70
duckweed 26
Ducula concinna (pigeon) 108
duku. *See* langsat
dukun (traditional healers) 68
durians 51, 53, 64, 83, 85, 96, 97, 97, 128, 129
Durio spp. (durian). *See* durians
  D. carinatus 64
  D. kutejensis (lai) 129
  D. zibanthinus 53
dye 71, 101, 106
Dysoxylum arborescens 26

E

eaglewood 71, 71
ebonies 6, 16
Echinochloa crus-galli (barnyard grass) 127
ecological disturbance 78–81
eelworm 113
eggplants 129
Eichhornia sp. (water hyacinth) 35
  E. crassipes 126, 126, 127
Eka Karya Gardens 134
Elaeis guineensis (oil palm) 98, 116–7, 117, 135
Elephas maximus (Asian elephant) 58
Elettaria (ginger species) 106
Ella Ulla River 63
emergents 12, 15, 16, 32
emping teki (local speciality) 127
endemism 9, 11, 24, 35, 83, 85
Endogonaceae 19
Endomycopsis (yeast) 95
Endospermum 60, 61
Entada rheedei 90

Entoloma sp. (fungus) 19
environmental destruction 78–81
epiphytes 16, 39, 60, 120, 132, 133
Epipogium roseum (orchid) 121
Equisetum ramosissimum (horsetail) 24
erosion, soil 79, 81, 90, 111
eruptions, volcanic 11, 52, 132
Erythrina sp. 91
Etlingera (*Nicolaia*) spp. (ginger) 106, 106
  E. elatior (torch ginger) 107
  E. solaris 16
Eucalyptus spp. 26, 31, 118, 133
  E. alba lily
Eugeissona utilis (sago palm) 98
Euphorbiaceae 26, 46, 60, 64
Euphorbia spp. (spurges)
Eurya 40
Eusideroxylon zwageri (Borneo ironwood) 77
export (diagram) 80, 83
export bans 72, 76
extinction 121, 123, 131

F

Fak Fak 58
ferns 6, 9, 16, 24–5, 35, 39, 60, 65, 88, 127
fertilisation 26, 29
fibre species 89
Ficus (figs). *See* figs
  F. benghalensis (banyan trees) 134–5
figs 26, 39, 53, 53, 64, 133
fishing 69
flame of Irian 31, 91, 124
flooded rice 86, 86
flora 10, 28–9, 64, 130–1, 131, 133
flower pecker 54, 55
flying foxes 51, 52
fodder 92, 93, 94, 95, 127
food source 64–5, 89, 92, 94, 98, 102, 103, 107
foot rot 111
forage 127
forests, types of 30–47, 80–1. *Also see individual listings*
  forest clearance 81, 89, 131
  forest fires 39, 81
  Borneo (1983) 46, 47
  forest land use 78–81
  forest management 118
  forest products 62–81
  forest regeneration 14, 46–7
  forestry 70–81 controlled 78–81
  fossil records 23, 24, 27, 84, 84
fragmentation, continental 10, 38
freshwater swamp 13, 34, 130, 132
Freycinetia spp. (climbing pandans) 16
fruits 46, 52–3, 64, 65, 83, 88, 96, 97
  tropical 129
  Also see citrus fruits
fruit pigeons 52
fruit tree islands 88
fungi 6, 9, 18–9, 20, 61, 95, 112
furniture industry 72, 91, 119
Fusarium moniliforme (crop disease) 18

G

gaharu 71, 71

galactomannan 93
gametophyte 23, 24–5
Gandaria (local fruit) 96
Ganoderma sp. (fungus) 19
gaplek 95
Garcinia mangostana (mangosteen) 16, 83, 85, 96, 97
gatot 95
gebang palms 44, 45, 98, 99
gedung 65, 65
gemmae 22
Gentiana spp. 39, 40, 41, 41
  G. lateriflora (gentians) 39
geocarpy 92
germ plasm 128
germination 55
Gigantochloa sp. 74
gingers 16, 16, 88, 106–7, 132
ginger lily 107
Ginkgo biloba (maidenhair tree) 27, 27
Ginkgoopsida (gingkos) 27
glacial periods 10–1
Gleicheniaceae 25, 25
Gliricidia sepium 93
Glycine max (soybean) 83, 92, 93, 97
glycogen 60
Gnetopsida 27
Gnetum spp. 69
  G. gnemon (meninjau) 97
Gondwana 10, 38
Gonocaryum seedling 53
Gonystylus spp. 119
  G. bancanus (ramin) 53
Gramineae 74
Grammatophyllum speciosum (leopard orchid) 120, 121
grasses 42, 44, 127
green beans 93
green pigeons 53
green revolution 87
groundnut 92
guar 92–3
guava 97
gula merah 102
Gunung Agung 86
Gunung Baluran 13, 43
Gunung Gede 16, 121, 133.
Gunung Gede-Pangrango National Park (Java) 16, 39, 40, 121, 133, 133
Gunung Jaya 40, 133
Gunung Kerinci National Park (Sumatra) 12, 130, 130–1
Gunung Leuser National Park (Sumatra) 132, 132
Gunung Palung 14, 32
Gunung Rantemario 40
Gunung Tujuh National Park (Sumatra) 38
gutta percha 64, 68
Guttiferae 64
gymnosperms 6, 27

H

Halophila ovalis (seagrasses) 37, 37
halophytes 36
Hanselula (yeast) 95
hardwoods 76
haustorium 54, 55
healers, traditional. *See dukun*
health centres 68
heath forest 13, 32–3, 32, 40, 61, 72, 132
Hedychium coronarium (ginger lily) 107
  var. *flavescans* 107
  var. *chrysoleucum* 107
Helixanthera xestophylla (mistletoe) 55
hemi-epiphytes 16, 17
hemi-parasites 17, 54

herbs 16, 42, 44, 69
herbaceous swamps 34, 35
herbicides 126
Heritiera simplicifolius 32
Hevea brasiliensis (rubber tree) 77, 116, 117, 119. *Also see* rubber cultivation
hibiscus 125
Hibiscus rosa-senensis L. 125
hill forest 13
Holarrhena antidysenterica 49
Holocene 11
honey 64
hornbills 52, 52
Hornstedtia (ginger) 106
hornworts 22, 23, 23
horsetails 24
hummingbirds 51
hybridisation 122, 123, 124
Hydnophytum (epiphytic ant plant) 60–1
Hydnum repandum (tooth fungus) 19
Hydrilla verticillata 35
Hygrophorus (fungi) 18
Hymenocallis littoralis (spider lilies) 125
Hymenophyllaceae (filmy ferns) 24–5, 25
hyperparasites 54
HYVS (high yielding varieties of rice) 87

I

Ice Age 43
ijuk (palm fibre) 102
ikat weaving 131
illepe nut 64
Impatiens sp. (balsam) 39
Imperata cylindrica (alang-alang) 10, 45, 126, 127, 127
Ipomoea pes-caprae (beach vegetation) 11
incense 71
Indian rosewood 43, 76, 91
indica (rice type) 86
Indonesian maple 39
introduced species 83
Intsia spp. 76, 91
  I. palembanica (merbau) 76
Ipomoea aquatica (kangkung) 127, 127
Ipomoea batatas (sweet potato) 85, 94
Irian Jaya 31, 32, 33, 36, 38, 40, 42, 61, 81, 83, 85, 120
ironwood 77
IRRI (International Rice Research Institute) 6, 87
irrigated rice 86, 86, 87
irrigation 86, 87
isdia (lichens) 21. *Also see* lichens
Isoetaceae (quillwort) 24
Isonandra palaquium (gutta percha) 68
Ixora javanica (Javanese Ixora) 50, 124, 125
Ixora timorensis 36

J

jackfruits 85, 96
jajagoan (barnyard grass) 127
Jambi 77
jamu 68, 69
jati. *See* teak
Java 31, 34, 36, 43, 86, 96, 109, 118, 121, 122
Javan long 111
Javan rhinoceros 132
Javanese edelweiss 40, 133, 133
Javanese Ixora 50, 124, 125
jelutung 34
jernang 71

141

Job's tears 46

K

Kacang Bogor (bambara groundnut) 92
Kaempfer, Engelbert 120
Kaempferia spp. (ginger) 106
   K. galanga (kencur) 107
Kalimantan 31, 32, 33, 34, 44, 64, 68, 76, 80, 81, 83, 86, 88, 96, 116, 128, 129
kangkung 127, 127
kapok tree 89
kapur 70
Karo Batak (Sumatra) 67
kauri. See dammar
kawitan 66
kayu putih 35, 42
kebun 88–9, 89
Kebun Raya, Bogor. *See* Bogor Botanic Gardens
kecubung (weed) 127
kemanga (mango) 96
kempas 76, 91
kenari nuts 85, 97, 105
kencur 107
Kenyah Dayak (Kalimantan) 68
kerangas. See heath forests
keranji 91
Kerinci Seblat National Park (Sumatra). *See* Gunung Kerinci National Park
keropok (local snack) 97
Komodo National Park (Lesser Sundas) 36, 42, 133, 133
Koompassia (kempas) 76, 91
   K. excelsa 76
kopi luak 53
Korthalsella sp. (mistletoe) 54
Korthlsella dacrydii 54
Krakatau. *See* Anak Krakatau
kretek 109, 109
Kubu (Sumatra) 65
kudzu 90
kunyit. See turmeric

L

lactic acid bacteria 95
lai (fruit) 129
lakes 35, 61, 130
Lake Air Hitam 34
Lake Discovery 40
Lake Maninjau 35
Lake Matano–Towuti 35
Lake Toba 11, 77, 132
Lampung 35, 94, 109, 111
land clearance. *See* forest clearance
land conversion 36
langsat 97, 128, 129
Lansium domesticum (langsat). *See* langsat
lantana 124, 133
laos. See lengkuas
Lapita Culture 85
Laurasia 10, 39
leafhoppers 87
Lecanopteris (ant plants) 60, 61
legumes. *See* Leguminaosae
Leguminaosae 26, 43, 60, 64, 90–3
Lejeunaeceae (liverworts) 23. *Also see* liverworts
lengkuas 107
Lentinus spp. (fungi) 18
Lepiota spp. (fungi) 18, 19
Lepironia articulata (sedges) 34
Leptochloa chinensis (timunan grass) 127
Lesser Sundas 120
Leucaena leucocephala 93
liana 16, 17, 90
lichens 8 , 20–1, 39
lignin 59

limestone forest 32, 33, 120
limited production forest 80
Limnocharis flava (sawah lettuce) 127
lise 66
Litchi chinesis 97
Lithocarpus (oaks) 39
liverworts 6, 9, 22, 22, 23
Livistona rotundifolia 98
Lobaria 20
Lobelia sumatrana (herb) 11
logging 13, 14–5, 35, 78, 78, 79, 80, 81, 131
lontar palm 44–5, 98, 99, 102, 102, 103, 133
Loranthaceae (mistletoes) 17, 54. *Also see* mistletoes
Lore Lindu National Park (Sulawesi) 133, 133
Lorentz Nature Reserve (Irian Jaya) 133
lotus 34
lower montane forest 12, 12
lowland rain forest 9, 12, 13, 13, 14, 16–7, 31, 32–3, 53, 64, 69, 70, 89, 99, 130, 132
Ludwigia adscendans 34
Lycopodiaceae (clubmosses) 24, 24
Lycoperdon perlatum (Puffball fungus) 18

M

Macaca nemestrina (pig-tailed macaques) 49, 59, 100, 101
Macaranga 46, 60, 61
   M. gigantifolia 47
mace 108
Macrolepiota spp. (fungi) 18
Macrosolen spp. 54, 55
   M. avenis 54
   M. beccarii 55
   M. formosus 55
Magnoliaceae 26, 40, 51
mahogany (timber) 83
maidenhair tree 27, 27
maize 46, 83, 85
Malayan tapir 49
Malesia 12, 24, 31, 64
Mallotus 46
Maluku 32, 42, 44, 83, 120
Mambai 66
manau (rattan) 119
Mangifera spp. 53
   M. casturi 65
   M. indica 96, 128, 129
mango 53, 64, 65, 83, 85, 96, 128, 129
mangosteen 16, 83, 85, 96, 97
mangrove forests 13, 31, 36–7, 49, 76, 120, 132, 133, 133
Manihot esculenta (cassava or tapioca) 94, 95
Manila copal 71, 76
maple, Indonesian 39
Marantaceae 16
Marasmius (fungi) 18
Marchantia (liverwort) 22
marigolds 35
Maros Hills 32
mata kucing (fruit) 129
matoa fruit 64
mat-making 34
medicinal properties 54, 63, 64, 68–9, 71, 97, 101, 107, 110, 127
medulla 21
Melaleuca spp. 35, 35, 42, 43, 44, 70, 133
   M. cajuputi (paperbark) 35, 35, 42, 43, 45
   M. leucodendron (paperbark) 70, 133
Melanesia 85, 100
Melia azedarach 43

Meloidogyne javanica (nematode) 111
mengkulang 77
meninjau 6, 97
Mentawai Islands 69, 83, 105
Merangin 29
meranti 76, 81
Meratus Mountains 44
Merauke area 44
merbau 76, 91
mersawa 76
Meru Betiri National Park 111
Meteorium (liverwort) 22
Metroxylon sagu (sago palm) 35, 58, 62, 64, 65, 83, 98, 98, 104–5
Michelia sp. (ambon flower) 26
   M. champaca (champak) 67
migration 46
milinggame 94
millet 66, 85
Mimosa sp. 90
Mimosoideae 90, 91
mining 41
minyak kayu putih 70
minyak keruing 70
Miocene epoch 10
mistletoes 17, 36, 54–5
mole crickets 113
monkeys 59
monkey puzzle 27, 76
monocotyledons 16
monsoon 13, 43
monsoon forests 32, 42–3, 130
montane forest 31, 32, 38–9, 61, 132, 133, 133, 135
Moraceae (figs) 16, 26, 64
Morowali Nature Reserve (Sulawesi) 133
mosses 6, 9, 22, 23, 23
moss forest 39, 40, 133
moths 51
   larvae 112, 117
moulds 19, 95
Mountain Garden, Tjibodas 134
Mucor (bread mould) 19, 95
Mucuna bennettii (flame of Irian) 31, 91, 124
mudflats 36
mungbean 93
Musa sp. (bananas) 16, 67, 128, 128, 129, 130. *Also see* bananas
mushrooms 18, 19, 127
Mycena (gilled fungus) 18
mycobiont 21
mycorrhizal fungi 19
Myrmecodia (epiphytic ant plant) 33, 60, 61, 61
Myristicaceae 16, 26
   Myristica fragrans (nutmeg) 30, 108–9, 108, 129
Myrtaceae (myrtles) 13

N

Najas marina ssp. *sumatrana* 35
nandu wood 91
national parks 132–3. *Also see individual listings*
native gardens 88–9, 89
nature reserves 80. *Also see individual listings*
nectar 50, 51, 60, 90
Nectariniidae (bird) 50
neem tree 43
Negara River swamps 35
Nelumbo nucifera (lotus) 34
nematode 111
Nepenthes sp. (pitchers) 56, 56
Nephelium spp. 64, 96, 97, 128, 128, 129
   Nephelium lappaceum (rambutan) 64, 96, 97,

128, 128, 129
Nephelium malaiense (*mata kucing*) 129
Nephelium ramboutan-ake (pulasan) 96, 97
Neurospora 19
Ngaju Dayak 66
nibong 98
Nicolaia (Etlinger) solaris (ginger) 16
Nicotiniana tabacum (tobacco) 83, 113, 113
Nile Cabbage 34, 35
nipa palm 98, 98, 99, 103, 103
Nothofagus spp. (beech trees) 38
Nusa Kambangan 71
Nusa Tenggara 31, 36, 42, 44
nuts 96-7
nutmeg 16, 26, 30, 83, 88, 108–9, 108, 129
nutsedge 126, 126, 127
Nyi Pohaci 66
Nymphaeaceae 26
Nymphaea nouchali (water lily) 35
Nypa fruticans (nipa) 12, 37, 98, 98, 99, 103, 103
Nypoideae 103

O

oaks 39, 43. *Also see* she-oaks *and* Ceylon oaks
Odoardo Beccari 61
Ogan–Komering swamps 35
oil palm 7, 83, 98, 98, 116-7, 117, 135
Olearia (daisy bushes) 41
Oncosperma tigillarium (nibong) 98
Oomycetes (fungi) 19
orang utan 49, 73, 132
orchids 8–9, 9, 16, 23, 25, 39, 50, 61, 91, 120–3, 131, 133
ornamental plants 83
Oryza spp. (rice) 26, 84, 85, 86
   O. glaberrima 86
   O. nivara 84
   O. japonica 86
   O. javanica 86
   O. sativa 26, 85, 86
Ottelia alismoides 35
Ottelia mesenterium 35

P

Paboya Reserve 42
Pachyrhizus erosus (bengkowang) 93
Padang Lawas 44, 89
paddy 86. *Also see* rice cultivation
paku 65
Palaeoclimate 10
palanar 71
palms 50, 58, 88, 98–105, 132
palm cabbage 65
palm wine 102
Palmae 64
palmyra palm. *See* lontar palm
Palu Bay 42, 44
pandan. 16. *See* pandanus
Pandanus spp. (pandan) 34, 62, 65, 99, 132
   P. furcatus 34
Pandanus moss forest 38–9
Pangaea 10
Panicum repens (torpedo grass) 127
papaya 46, 96, 97
paper pulp 77, 80
paperbark tree 35, 35, 42, 43, 45, 70, 133
Paphiopedilum spp. (slipper orchid) 120, 121, 122
   P. praestans 9
   P. urbanianum 9

Papilio memnon (butterfly) 50
Papilionoideae (legumes) 90, 91
Papuasia 98
Paraphalenopsis laycockil (orchid species) 121, 121
Paraserianthes (batai) 91
parasitic plants 16, 18–9, 28
Parkia speciosa (stinking beans) 92, 93, 97
Pasemah 44
Paspalum conjugatum (buffalo grass) 127
Passiflora edulis (passion fruit) 96
passion fruit 96
patch canker 129
peanuts 92
peas 90
peat swamp forest 13, 34, 61, 122, 132
Pediocaccus (lactic acid bacteria) 95
peg 92, 93, 93
Penan (Borneo) 98
peppers 26, 83, 85, 110–1, 110, 111
perennials 44
Pericopsis (*nandu* wood) 91
perithecia 21
Perlindungan Hutan dan Pelestarian Alam 81
Peronema canescens (sungkai) 119
Persea americana (avocado) 97
pests, plant 87, 109, 111, 112, 113, 115, 116
petai (stinking beans) 51, 83, 92, 93, 97
Peziza sp. (cup fungus) 18, 19
Phalaenopsis spp. (orchid) 120, 122, 122, 123
   P. amabilis 9
   P. javanica 121
Phaseolus vulgaris (green beans) 93
photobiont 20, 21
photosynthesis 16, 20, 21, 54, 58
Phylidris sp. 60
   P. cordatus (ants) 60–1
phyllidia (lichens propagules) 20, 21
Phytophthora (crop disease) 18
   P. palmivora (foot rot) 111
pigeon 108
pinang sirih. See Areca catechu
pine 27. *Also see pinus merkusii*
pineapple 96
Pinopsida (conifers) 27
Pinus merkusii (Sumatran pine) 27, 43, 70–1, 77, 118
Piper spp. 52, 67, 101, 101, 111, 111
   P. aduncum 52
   P. betle (betel leaf) 67, 101, 111
   P. nigrum (pepper) 110–1, 110, 111
   P. retrofractum (javan long) 111
pipeworts 56
pisang. See bananas
Pistia stratiotes (Nile Cabbage) 34, 35
Pisum sativum (pea) 90
pitcher plants 33, 39, 56–7, 56, 57, 132, 133
plantain squirrel 48–9
planthoppers 87
Platycerium (stag's horn fern) 25
Plectocomia sp. 72
Pleistocene Ice Age 42
Pleurotus sajor–caju

(oyster mushroom) 19
Plocoglottis lowii (orchid) 121
plywood 76, 80, 118
pneumatophores (air roots) 34, 35, 36, 132
Podocarpus 39
  P. neriifolius (conifer) 11
poisons 65, 69
pokah bung 18
pollen grains 50
pollination 6, 14, 17, 28, 29, 49, 50–1, 53, 54, 77, 90, 121
pomelo 97
Pometia pinnata (matoa fruit) 64
potatoes 83, 94
Potentilla sp. 41
  P. borneenis 11
preservation, forest 78–81
primary forests 65
production forests 80
propagules 21, 22, 52, 123
protection forest 80
Psidium guajava (guava) 97
Psophocarpus tetragonolobus (climbing winged bean) 93
pteridophytes 6, 24
Pterocarpus spp. (timber) 91, 118
  P. indicus 76
Pteropus (flying fox) 51
Pueraria phaseoloides (kudzu) 90
puffball fungus 18
pulasan 96, 97
pulpwood 77
Purwodadi Botanic Gardens 134
puskesmas 68
puspa langka 29
Pyricularia oryzac (crop disease) 18
Pythium aphanidermatum (crop disease) 18

Q

quillworts 24

R

Rafflesia 6, 9, 17, 17, 50, 132
  R. achensis 30
  R. arnoldii 28–9, 29, 130, 131
  R. hasseltii 29
rain forests 8, 9, 14
rain tree 91, 91
rainfall 13
rainfed rice 86, 86
Ramaria spp. (coral fungi) 18
rambutan 64, 83, 96, 97, 128, 128, 129
ramin 53, 119
Ranau 44
Ranunculus 40, 41
rapu 66
rats 112
rattan 6, 16, 60, 62–3, 63, 64, 71, 72–3, 98, 99, 119, 132
Rawa Danau Reserve 34
  rebung (bamboo shoots) 63, 64, 75
regular production forest 80
Renanthera sp. (orchids) 122
reproduction process 50
  asexual 19, 20, 21, 22, 24
  fungi 19
  ferns 24
  lichens 20, 21
  liverwort 24
  sexual 22–3, 24
  vegetative 22, 126
resin 63, 64, 70–1, 88, 118
rhizoids 17
rhizomes 60, 74, 106–7, 126

Rhizophora sp. (timber) 36, 36, 37, 37, 76, 119
  R. sapling 26
Rhizopus spp. (bread mould) 95
  R. oligosporus 19
rhododendrons 39, 40, 41, 41, 120, 132
  R. javanicum 6
  R. saxifragoides 41
Rhodomyrtus tomentosa 26
Rhynocophorus ferrugineus (weevil) 104, 105
rice cultivation 26, 46, 66–7, 83, 82–3, 84–7, 129.
rice bugs 87
rice ecosystems, types of 86, 86–7
Ridley, H.N. (botanist) 116
riedelia sp. 107
Rimbo Kilm 77
rituals 66–7
root crops 64, 94, 95
root parasite 16
rootstock 129
rose apple 51
rosewood 76
Roti 44, 98
rubber cultivation 77, 82, 83, 88, 88, 89, 116, 117, 119 *Also see* Hevea brasiliensis
Rubiaceae (ant plant) 60
Rumphius, Georgius Everhardus 10
Ruteng Nature Recreation Park (Lesser Sundas) 133

S

Saccharomyces (yeast) 95
Saccharum sp. (sugarcane) 112, 128
sago palms 35, 58, 62, 64, 65, 83, 98, 98, 104–5
Sahul shelf 10, 42, 43, 44
Salacca edulis (zalacca) (salak) 58, 96, 97
Salvinia molesta (water fern) 35, 127
Samanea (Albizia) saman (rain tree) 91
sandalwood 6, 43, 91, 130
sandat 67
Santalum album (sandalwood) 42, 43, 130
Sapindaceae 64
saponins 65
Saprophytes 17
Saracca (legume) 91
savanna grasslands 43, 44–5, 102, 133, 133
Savu 44, 66
sawah lettuce 127
sawmilling 80
Scaphium (mengkulang) 77
Schizostachyum 74
Schleichera oleosa (Ceylon oak) 43
Scytinopogon angulisporus 18
sea levels, changing 10, 42
seagrasses 36, 37, 37
sealing wax palms 6, 99, 124, 125, 131
seasonal forests 13
secondary forest 46–7, 60
secondary succession 46–7
sedges 34, 35, 40, 56
seed dispersal 6, 49, 52–3, 54, 55, 108. *Also see* dispersal agents
Sekonyer river 132
Selaginellaceae (clubmosses) 24
Selaginella sp. 24
self pollination 90. *Also see* pollination
sengon 91
sepetir 76, 91
seram 16, 44
serpentine forest 33, 120

Sesbania grandiflora 93
Setaria italica (millet) 85
sexual reproduction. *See* reproduction
she-oak 11, 39, 43, 45, 50, 133
shifting cultivation. *See* swidden cultivation
Shorea spp. (meranti) 76
  S. curtisii (Dyer ex King) 33
  S. javanica (dammar tree) 70
  S. macrantha (illepe nut) 64
shrubs 40
Siberut Island 65, 71
Sindora (sepetir) 76, 91
slash-and-burn agriculture. *See* burning forest
snails 111
snakes (plant pest) 112
snakefruit palm 58
softwoods 76, 118–9
soil erosion. *See* erosion
Solanum melongena (eggplant) 129
Sonchus arvensis (corn sow thistle) 127
Sonneratia spp. 51
  S. alba 37, 37
  S. caseolaris 37
soybeans 83, 92, 93, 97
spanish needles 126
spice trade 108
spider lilies 125
Spiridens (moss) 23
sporophyte 23, (diagram) 24
spurges 26
squash 46
stag's horn fern 25
starfruits 96, 97
stem borers 87, 112
Stevia rebaudiana (stevia shrub) 113
Sticta (lichen) 20
stilt roots 34
stinking beans. *See* petai
strangler fig 17
stratification, forest 14, 15
strychnine 58, 69
Styphelia (Australian heath) 40, 41
Styrax benzoin 70, 71, 88
sub-alpine rain forest 12, 12, 132, 133
sub montane forest 133
sugarcane 66, 83, 88, 95, 112-3, 112, 128
sugar palm 98, 102, 102
Sulawesi 32, 33, 44, 61, 80, 81, 86, 109, 128, 118, 120
Sumatra 31, 34, 43, 44, 73, 76, 80, 81, 83, 86, 88, 89, 109, 111, 115, 116, 118, 121, 132
Sumatran pine 27, 43, 70–1, 77, 118
Sumatran rhinoceros 58
Sumba 130
sunbird 50, 51
Sunda shelf 10, 44
sundews 56, 57, 57
sungkai 119
swamp forest 6, 31, 34–5, 132, 133
sweet potatoes 83, 85, 94–5, 94
swidden cultivation 46, 47, 65, 83, 85
Syzygium spp. 34
  S. aromaticum (cloves) 83, 109, 109, 129
  S. javanica (waterapple) 97

T

Tabernaemontana spp. 69
Tacca integrifolia (herb) 69
Talang Mamak (Sumatra) 68
tamarind 43, 44, 92, 93
Tamarindus indica

(tamarind). *See* tamarind
Tanah Merah 65
Tapeinocheilos pungens (ginger) 7
Tangkoko Batuangus Nature Reserve (Sulawesi) 133
Tanimbar 44
Tanjung Pinang 88
Tanjung Puting National Park (Kalimantan) 32, 132, 132
tannins 59, 65
tape (fermented block cassava) 95
tapioca. *See* cassava
Tapirus indicus (Malayan tapir) 49
taro 66, 85, 85, 94, 95
tea 40, 83, 114-5, 114, 115
teak 43, 77, 83, 118, 119
Tectona grandis (jati) 43, 77, 77, 118, 119 see teak
tempe (fermented soybean) 97
Tengger 11
terranes 10
Tethys Ocean 10
Tetrastigma (liana) 28
Tetun 66
thatch 102, 103, 105, 127
Thelypteridaceae (ferns) 25
Themeda (grasses) 44
Theobroma cacao (cocoa) 115, 115
thorn forest 42, 58
Tiliaceae 64
timber 6, 32, 39, 45, 49, 64, 70, 71, 72, 76–7, 78–9, 79, 91, 118–9. *Also see* Dipterocarpaceae
timber industry 46, 63
Timor 44, 85
titan arum 6, 28–9, 28, 29
tiwul 95
tobacco 83, 113, 113
tomatoes 85
Toraja (Sulawesi) 66
tourism 39, 41
toxins 58, 65
traditional cures 63, 64, 68–9
Trangan 44
transmigration programmes 46, 81
tree ferns 135
tree kangaroo 58
tree of life 66
treelets 15, 16
Treron sp. (green pigeons) 53
tropical rain forest 42, 132, 132
true grasses 35
Truebia (bryophytes) 23
tuak (palm wine) 102, 103
tuba 65, 69
tubers 64
tumpangsari system 119
tumpeng robiyok 67
tundra 40
turmeric 106
turpentine 70
tussock grassland 40
tuwuhan 67

U

Ujung Kulon National Park (Java) 132–3
Ulu Leang cave 84
understorey vegetation 42, 46, 69, 132
upas tree 69
upper montane forest 12, 12, 120
Utricularia spp. (bladderworts) 56–7, 56
U. aureum 35
U. gibba 56

U. uliginosa 56

V

Vaccinium 40
van Steenis (botanist) 42, 43
Vanda sp. (orchids) 9, 122, 122
Vanilla spp. (orchid) 120, 121
  Vanilla pilifera 121
Vavilov Centres 129
vegetables 83, 85, 88, 93
vegetation, types of 12, (map) 30–1
vegetative reproduction 22, 126
Vigna spp. 92, 93
  V. radiata (mungbean) 93
  V. subterranea (kachang bogor) 92
  V. unguiculata (yard-long bean) 92, 93
Viola (violets) 39, 40
Viscaeae 54
Viscum album (European mistletoe) 54
volcanic eruptions 11, 52, 132
Volvariella volvacea (padi straw mushroom) 19, 127

W

Wallace, Alfred Russel 10, 53
Wallace'sline (diagram) 10–1
Wanariset 11
waringin trees 66
Wasur National Park (Irian Jaya) 133, 133
water convolulus. *See* kangkung
water dispersal. *See* dispersal agents
water fern 35, 127
water hyacinth 35, 126, 126, 127
water lettuce 34
water lily 35, 51
waterapple 97, 97
watermelon 96, 97
Way Kambas National Park 12, 46, 132
weeds 25, 126–7
weevil 104, 105, 111
Weismann, August 128
wet rice cultivation 85
Wetar 43
wetlands 34–5, 36–7, 36
whitewood 42
wild foods 65
wild meritus 97
wild plant foods 64
wind dispersal. *See* dispersal agents
Winteraceae 26
Wolffia 9
woodcarvings 43, 75
woodpulp 71, 76, 80
World Conservation Union (IUCN) 78
worms 117

X

Xylia xylocarpa 76

Y

yams 26, 65, 66, 83, 85, 94, 95
yard-long bean 92, 93
yeasts 18, 19, 95

Z

Zingiberaceae 16
Zingiber officinale (ginger) 107, 107
Ziziphus mauritiana 44
Zyziphus trees 133
zygomorph flowers 90
Zygomycetes (fungi) 19, 95

Photo Credits

Unless otherwise specified, all the pictures and illustrations have been supplied by Editions Didier Millet.
Chapter openers: Diversity of Indonesian Flora, p.8, diversity of rain forest, (6), David Stone; orchids, Atang Fachruroji. Vegetation types and Biogeography, p.30, flora (4), Alain Compost; Plant–Animal Interactions, p.48–49, animals/plants, Alain Compost, bamboo, Bruce Coleman. Forest Produce, p.62–63, (3), Alain Compost. Cultivation, p.82–83, rice terraces, Photobank; Conservation, p.130–131, Indonesian flora (5), Alain Compost.
Reproduced by kind permission of: The Natural History Museum, (NHM), London, p.24, clubmoss; p. 25, *Hymenophyllum*. IRRI, p.87, organisms.

The publishers have made every effort to ensure that all photographs and illustrations contained in this volume have been correctly credited, and apologise if any omissions or errors have occurred. These will be amended in the next printing.

Antiques of the Orient, p.58, vogelkop tree kangroo; p.68–69, gutta percha; plants as poisons; p.70, *S. benzoin*; p.94, roots/tubers; p.97, candle nut; p.99, Nibong; p.100, pepper harvest. p.106–107, turmeric, commercial ginger; p.108–109, map; p.110–111, Pedie, pepper harvest; p.112–113, 19th c. sugar factory, tobacco cultivation, sorting, and transportation; p.115, tea; p.134–135, avenue of Waringins, Bogor Botanic Gardens (3).
Deni Bown, p.28–29, Rupiah, Arum lilies.
Julian Caldecott, p. 79, logging trails.
Martin Cheek, p.56–57, *U.gibba, U. uliginosa, D. peltata*.
Janet Cochrane, p.133, mangrove, termites.
Bruce Coleman, p.10, *Barringtonia*; p.18–19, puffball, fungi (A); p.24, tree fern; p.32, limestone cliffs; p.43, monsoon; p.46, secondary forest; p.51, *Sonneratia*; p.53, figs; p.57, *D. indica*; p.71, camphor trees; p.94, sweet potato; p.111, pepper plantation.
Alain Compost, p.7, oil palm; p.10–11, Anak Krakatau, alang-anang, conifer; p.12, *N. fruticans*; p.15 logging road; p.16–17, *B. elongata*, ginger, cauliflory, liana; p.19, *Entoloma* (B); hairy cap fungus (G); p.24–25, *Selaginella; Gleichenia*; p.26, chempaka, *R. t tomentosa*, primitive angiosperms; p.28–29, *A. titanum*, survival (2); p.32, heath forest; p.35, paper bark tree (2); p.37, propgule,(3d); p.38–39, pandanus, *Impatiens*; p.40–41, edelweiss, alpine meadow; p.43, *Acacia*, Sumba craftsman; p.45, alang-alang; p.46–47, shifting cultivation, Borneo fire; p.50, bee, sunbird; p.52–53, dispersal by wind and water (3), fruit bat, hornbill, orang utan, coffee; p.58–59, *Metroxylon*, Salak, tapir, rhinoceros, macaque; p. 60, ants' nests (2); p.64– 65, Dayak, paku;p.68–69, poster, sachets, woman, jamu seller, Dayak, extraction of poison; p.71, resin, tapping gum, *Agathis*; p.72, Dayak; p.74,Toraja house; p.77, ironwood flowers; p.79, transporting logs, heavy erosion; p.81, cleared rainforests, tractor, muddy rivers; p.85, Sulawesi farmers, drainage ditches (2); p.86, cutting paddy; p.98, cinnamon; p.92, soyabean; p.94–95, tapioca processing; p.98, hen nest; p.100–101, macaques picking coconut, man with stained teeth; p.104–105, Asmat, sago tree ceremony; 108–109, clove plantation, harvesting, drying cloves; p. 112–113, crushing sugarcane, workers taking a break; p.114–115, coffee berries, tea plantation, Teh botol; p.117, tapping rubber, latex processing; p.118–119, transplanting, teak (3), timber; p.123, black orchid, production (6); p.125, bonsai; p.132–133, rainforest; montane forest; p.134, gates.
M.Coode, p.106–107, *C. speciosus, R.corallina*.
Jill Cowley, p.106, *C.globosus*.
John Dransfield, p. 60, rattan; p.102, sugar palm.
Soejatmi Dransfield, p.74, detailed weave.
Alan Eddy, p.22, liverwort, moss; p.23, *Calymperes,* moss forest, *Spiridens*.
Ian Edwards, p.53, Gonocaryon; p.91, Flame of Irian.
Jean-Paul Ferrero/ AUSCAPE, p. 59, proboscis monkey;
Wim Giesen, p.34, sedge.
Ian Glover, p.84, Ulu Leang 1, (4).
Rio Helmi, p.67, offering, areca nut; p.69, herbalist; p.100, husk; p.107, rhizome, p.116, plantation.
Hew Choy Sin, p.123, *D.nobile* (C).
Leo Hux, p.17, rafflesia.
Sian E. Jay, p. 88, home garden.
Mathew Jebb, p.61, *M. arfakianum*.
A. Clive Jenmy, p.24, horsetail.
Don Kirkup, p.55, *H. xestophylla, M. beccarii*.
Y. Laumonier, p.29, *R. hasseltii*.
G.P. Lewis, p. 106, *E. littoralis*.
David Mabberley, p.27, angiosperms, *Ginko bilolba*, cycads (2).
L.J.G. van der Maesen, p.76, timber types; p.90–91, *D. regia, C. pulcherrima, A. farnesiana, P. sativum*, raintree, legume leaves.
Genevieve Michon, p.98–99, resin collector, biodiversity.
John Miksic, p.127, alang-alang.
Kathryn Monk, p.42–43, ant plants, *Acacia;* p.45, savannas.
Photobank, p. 6–7, *cili* dolls; p.72, unloading rattan; p.76, paper pulping ; p.78, timber trucks, clearing primary forest, processing timber; p.96, fruit production; p.103, lontar buckets; p.122, *Vanda*; p.124, hibiscus, ornamental gardens (2).
Ivan Polunin, p.37, *Enhalus*, primary stem (3c); p.50, Ixora; p.76, *K. excelsa*; p.81, destruction; p.86, holy water; p.99, *gebang*; p.102, lontar reliefs; p.119, tambak; p.121, *B. lobbii*.
Øyvind Sandbukt, p. 65, grating *gedung*.
Jurg Schneider, p.94–95, Dani, slope cultivation.
Trond Schumacher, p.69, *T. integrifolia*, p.102, lontar manuscripts.
Mohamad Soerjani, p.126, barnyard grass.
Tara Sosrowardoyo, p.101, betel cutter.
David Stone, p.36, rhizophora; p.64–65 sago, pandanus; p.74, scaffolding; p.114, tea plantation.
Subandi, p.94, selling cassava.
T.K.Tan, p.18-19, fungi: pathogenic fungi; fungi – cap (C), tooth (E), bracket (F), stinkhorn (H).
H.Taylor, NHM, p.20, *filamentous* lichens, l. with crustose.
Harry Taylor, p.25, treetop gardens.
Alan Watson, p.14, *calophyllum*; p.17, herb; p. 21, lichen; p.32–33, *H. simplicifolius*; p.35, pneumatophore; p.36, mangrove; p.38–39, lichen, fern.
Tim Whitmore, p.47, *macaranga;* p.118, kauri.
A. J. Whitten, p.11, herb, beach vegetation, *P. borneenis*, she-oak; p.19, mushrooms; p. 28, male/female flower; p.34, swamp forest; p.41, leaf; p.57, red sundew; p.59, caterpillars; p.71, resin oozing; p.72, rattan; p.77, ironwood, teak flowers; p.80, sawmill; p.98, *Livistona*; p.102, sugar palm thatch, p.104–105, sago worms, sago containers (2); p.107, ginger lily; p. 110, pepper vines.; p.133, edelweiss.
Elizabeth Widjaja, p.129, vendor with eggplants.
Wildlife Matters, p. 17, bird's nest fern; p.59, butterfly; p.89, *kapok*; p.126–127, hyacinth, kangkung.

Illustrations by:
Anuar B. Abdul Rahim, p.14, inorganic nutrient; p.19, mycorrhizal fungus; p.25, life cycle; p.29, *Rafflesia* (5); p.35, peat swamp; p.36-37, root systems, seagrass; p.40, changes; p.46–47, *M. gigantifolia*; p.50-51, Rose Apple; p.54–55, flower pecker, spider hunter; p.56–57, pitcher trap, bladderwort; p.66–67, Tree of Life, *Tumpeng Robiyok*; p.77, teak; p.85, Dayak; p.86–87, ecosystems; p.116–117, stages, rubber tree; p.126-127, Purple Nutsedge.
Komang Agus Bagiada, p. 21, lichens;
Atang Fachruroji, p.118, chair; p.122–123, orchids (6); p.124-125, Flame of Irian, ornamentals (1–5)
Bruce Granquist, p.44–45, savanna types; p.96, fruits (B & E); p.112, sugarcane; p.127, alang–alang.
Mubinas Hanafi, p.74–75, culms, giant bamboo; p.96, snakefruit, (C); p.103, *N. fruticans*; p. 108, clove; p.120, parts; p.129, langsat (A–C).
Nengah Nurenten, p. 53, mango; p. 64, rambutan; p.101, areca nut, (1–4 & A–F); p.114, coffee.
Studio Satumata. p.12-13, changes with altitude; p.14-15, forest dynamics; p.73, versatile rattan (6); p.75, details of giant bamboo,(A, B & C); p.90–91, pods; p.93, peanut (2), p.95, cassava shrub; p.97, fruits 1–10, cashew nut; p. 108, mace; p.113, plant; p.114, branch; p.120, leopard orchid; p.124–125, ornamentals (A–D); p.128, rambutan (1–3), bananas (A–C).
Ujang Suherman, p. 92–93, tamarind, winged bean (A,B,1,2); p. 96, durian (A), dudu (D); p.9899, palms (1–10); p.127, slippers.
Susilowati, p.92, petai; p.99, fruit; p.103, inflorescence (2); p.115, pod.
Teo Eng Hean, p.20, lichens; p.21, lichen on leaves; p.22–23, life cycle of moss, *Calymperes sp.*, hornwort sporophyte; p.27, Sumataran pine; p.33, Dyer ex King; p.60–61, *Hydnophytum*; p.70–71, Dayak; p.84–85, spades, foxtail millet; p.98, farmers; p.110, harvesting pepper.

Kind assistance rendered by the following institutions and persons:
Bogor Botanic Gardens; Indonesian Forestry Community; International Centre for Research in Forestry, Indonesia; International Rice Research Institute, Philippines; National Parks Board, Singapore; Wetlands International, Indonesia; Ivan Polunin; Ian M. Turner; Christiana Soh and Zakiah Agil (Singapore Botanic Gardens Library); and Helen West (Antiques of the Orient).

9789813018297.2

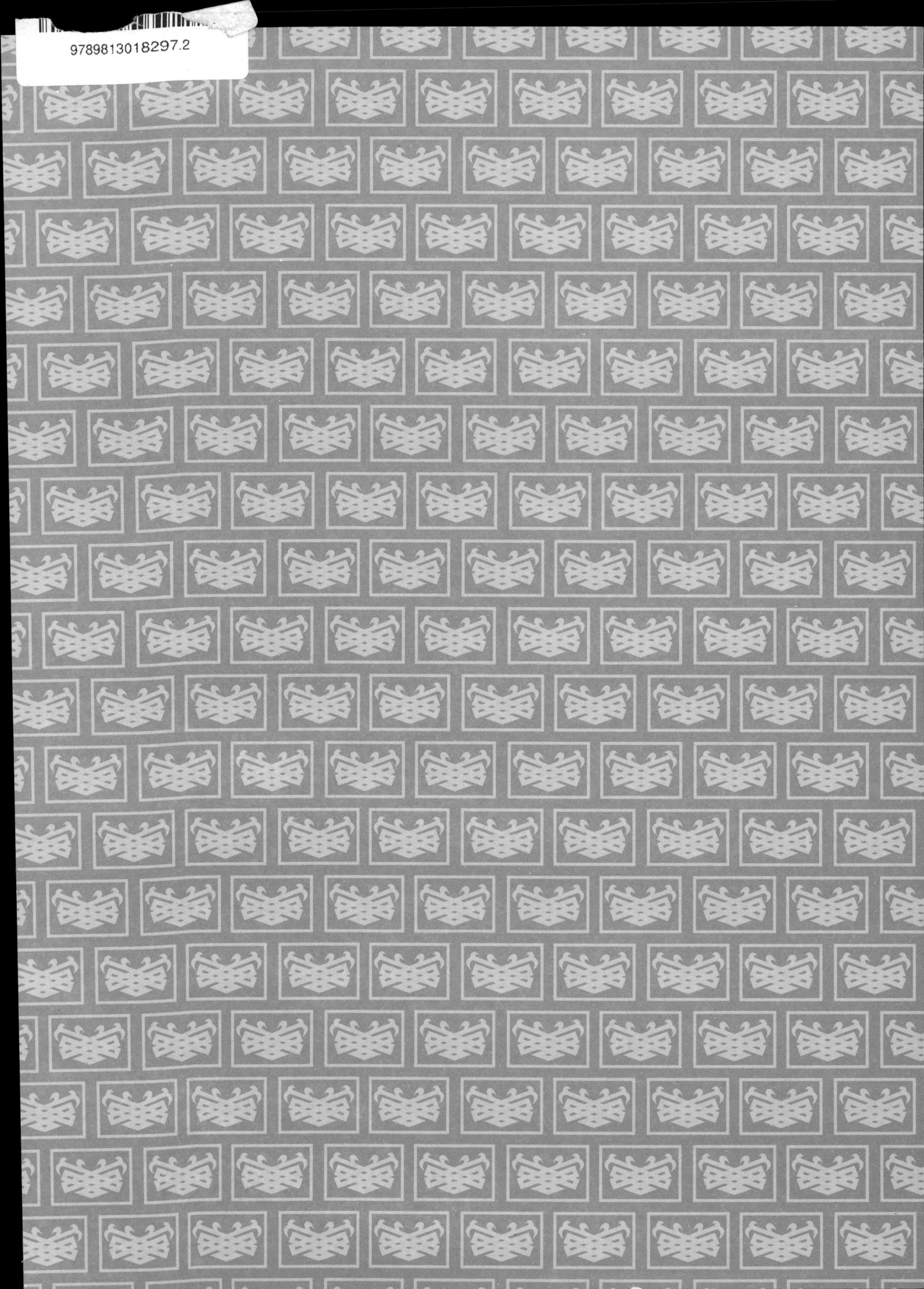